Study Guide

for

CHEMISTRY
A Contemporary Approach
Second Edition

Study Guide

for

CHEMISTRY
A Contemporary Approach
Second Edition

by G. Tyler Miller, Jr., David G. Lygre, & Wesley D. Smith

Wesley D. Smith
Ricks College

David G. Lygre
Central Washington University

Noel Zaugg
Ricks College

Wadsworth Publishing Company
Belmont, California
A Division of Wadsworth, Inc.

ISBN 0-534-07201-1 49

Printed in the United States of America

1 2 3 4 5 6 7 8 9 10---91 90 89 88 87

CONTENTS

INTRODUCTION
HOW AND WHY TO USE THIS BOOK

With a good textbook, why do you need a study guide? You'd think that a decent textbook by itself ought to cover all the necessary subject matter for the course. And, unless the book was organized in a cement mixer, you'd expect that starting with the first chapter and ending with the last is pretty much all the advice a study guide could give. So what exactly is this extra volume good for?

In an ideal world of perfect textbooks and perfect students no study guides would be necessary—but neither would colleges and universities. All students would simply obtain the appropriate books and educate themselves. The real world, however, possesses few such superhuman students and absolutely no flawless textbooks. Nevertheless, as a regular human being in an imperfect environment, you must still learn what you need to know on your own. No one can inject you with knowledge; you have to digest it yourself to make it a part of you. To help, universities have general education requirements, and professors teach courses so that you can gain knowledge more efficiently. But it's still up to you.

This study guide exists to increase your learning efficiency even more. It can give you several advantages that the textbook can't:

Informality The narrative style of a textbook, like that of a radio broadcast, tends to put some distance between the author and the reader. But here, I don't need to be so reserved. I can talk to you person-to-person as if we were on the telephone, and I can ask you to do specific things that I couldn't request in the textbook.

Structure The textbook authors (with my help) put a lot of thought into how the textbook should be organized, and they tried to make that plain in their writing. To be doubly sure that you'll benefit by their work, however, I'll give you an explicit outline of each section's organization. I'll also give you a complete set of objectives, which are often more pointed and performance-oriented than the questions at the beginning of each chapter, so that you'll know exactly what skills you'll need to master.

Drill and Practice Although a number of Practice Exercises are scattered throughout the text, from time to time you'll almost certainly feel the need for more. The urge will be strongest when confidence in your understanding is weakest. I'll supply you with plenty of worked-out examples in this study guide—enough to restore your confidence when it seems to be flagging.

Self Tests You'll always learn from your experience. But the textbook can't give you any feedback on your learning, and profiting from mistakes on midterms is costly. Thus, I'll test you over the material and suggest from the results what parts of it you might want to restudy. That way your experience will not come as a penalty, and you'll be able to perform better when it counts.

Even with a study guide, what hope do regular students have in a chemistry class? Chemistry is frightening because people perceive it as one of the most difficult of all college subjects. No doubt you've heard horror stories from friends or relatives about how hard chemistry is. In fact, some of the may even have flunked college chemistry. And if this happens to other reasonably bright people, you may be thinking you don't have a chance.

Don't jump to that conclusion. Chemistry is not the impossible dream, nor is it the sole property of science majors. Anybody with ordinary common sense can understand it, and, therefore, *you* can do chemistry too. It may not come easy, but you can allow for that. You do difficult things all the time. And deciding to do them in the first place is much more important than figuring out how. Thus, if you are determined to conquer chemistry, you can.

Pause here a moment, and make a commitment to yourself—in whatever way works best for you—that you will do what it takes to succeed in chemistry.

But what if you just don't care about chemistry? Over the past 15 years, I've asked all the students in my courses on Chemistry for Nonscience Majors the same initial question: "Which of you would *not* have taken chemistry if it weren't part of some college or university requirement?" The overwhelming majority of them raised their hands, so I'll bet you're in the same situation. That's okay; we all have different interests. I'd have bypassed the Shakespeare and German courses I had to take. But the fact is you're now enrolled in this chemistry course. How well you do, like it or not, will be your choice and yours alone. If you rebel and vow to do the minimum amount of work to get by, that's what will happen: You'll just get by. If you decide to learn all you can while you're at it, you will learn.

Here's a crucial question, whose answer will determine the direction you take. For the moment, forget about the requirements that forced you into this course, and disregard any noble ideas about what you "should" or "ought to" be doing. Simply find as many positive answers as you can to the question "What value will a general knowledge of chemistry give to my life in the years to come?" Write your answers here:

Are you reading this without having seriously answered the question on the previous page? Did you skip making the earlier commitment? If so, you're losing part of the value of this study guide already, and you're giving yourself early warning signals about what your future attitudes will be.

Having committed yourself to do well in chemistry, and having decided what good that success will do for you, you have taken the two most important steps. What's left now are the mechanics.

What skills do you need to understand chemistry? To learn chemistry (or any other subject), you must be able to (1) recall memorized material, (2) classify things into categories, (3) use rules, and (4) create rules. That's all; no other special abilities are necessary. And there are ways to become good at each of the four skills.

[M] **Recall** is the ability to remember the correct response to a given stimulus. It has to do with association—words and their meanings, events and their dates, states and their capitals, and so on. If everything is paired up properly in your mind, then when I say one half of the pair, you can immediately say the other.

Perhaps you've noticed how much of your grade in many classes depends on this skill. It's important in chemistry, too, but it usually carries a lesser emphasis. Mostly you'll need to recollect things only as a prerequisite to the other three skills. So I don't want you just to develop a facility for regurgitating or parroting things; I want you to get all the necessary material on the tip of your tongue so that you can use it later.

Some things take talent or brains, but memorizing just takes work. For most people, recall requires lots of repetition. Thus, if you want something to remain in your memory, see it, hear it, write it, and say it many times. The idea is to repeat, rehearse, restate, reiterate, recapitulate, and rerun it redundantly. I can't help you much with this chore; you have to do it on your own.

A great way to assist yourself is to create *mnemonics*, or memory aids. If you can imagine some elaborate and silly scene that connects a stimulus with the correct response, then the connection is yours forever. For example, if you wanted to recall the chemical symbol for sodium, you might visualize a little brat swimming in a huge glass of *soda* pop yelling "Na Na Na N' Na Na, betcha can't remember!" The more absurd you make it, the more memorable it will be. I've suggested some mnemonics here and there in the study guide, but the most effective ones will be those you make up yourself. So again, it's up to you to do the work.

I *can* help you see if the work you have done is sufficient, however. For each recall objective—marked with the symbol **[M]** for Memory—I've included a set of flash cards. Printed on one side is the stimulus; on the other is the response. Cut them out and keep them in your pocket or purse. Then go over them every time you have an idle moment. Put the ones you know in a separate pile, and continue to work with the ones you don't.

[C] **Classification** is the ability to categorize objects, events, or ideas into their correct groups. It has to do with the understanding of concepts—what is a dog? an acid? a letter A? an aromatic aldehyde? If you really comprehend a particular concept, I can show you any object, and you can tell me whether it belongs in the category or not.

Classification is a bit more difficult than recall because it goes beyond memory. First, you need to remember the definition of a concept; that is, you must know the critical characteristics that make something belong to the category. Then, you have to apply that knowledge to a given object, possibly to one you have never seen before. For example, memorizing that a bird is an animal with wings and feathers is only half the skill; the rest is knowing one when you see it. You must be able to identify as birds such exotic species as California condors, Australian emus, or Guatemalan quetzals, even though you may never have encountered them before.

For each classification objective—marked with the symbol **[C]**—I've given you a complete definition, underlining all the attributes that make a difference. You'll need to memorize each of these. Then, to help you distinguish between these critical attributes and the ones that don't matter, I've given you several examples of things that belong to the category, and I've matched them with things that don't belong but are similar in other ways. For example, a mud hen and a muskrat are both about the same size and color. Neither one can fly very well. Yet only the mud hen has wings and feathers, and only it is the bird.

I'll test your ability to classify by asking you to categorize wide and varied groups of instances.

[R] **Rule Using** is the ability to follow a recipe or carry out a procedure. It's understanding how to do something. If you have this skill, I can give you a task and you can complete all the necessary steps to accomplish it.

Rule using involves the previous two skills. Whenever you're presented with a problem, you have to classify it to determine what rule applies. Then you need to recall the procedure that will give you the answer. For example, if you have a flat tire, does it mean you use your tire-changing skills or your getting-someone-to-do-you-a-big-favor ability? If you decide on the former, do you jack up the car first and then loosen the lug nuts, or is it the other way around? Making the proper decisions is what leads to successful rule using.

For each rule-using objective—marked with an **[R]**—I've given you instructions either in the form of a numbered list or as a flow chart. A list gives you straightforward, one-after-another directions for a simple routine. A flow chart is a step-by-step procedure that includes any decisions you may need to make and what to do in each case. For instance, to make a call on a pay phone, you might use this procedure:

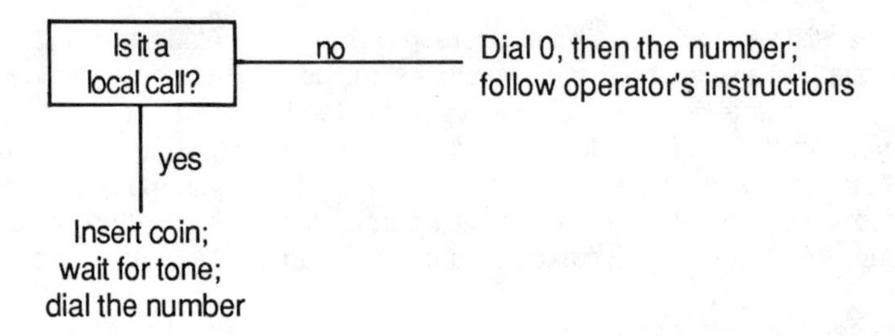

Lists and flow charts make it much easier for you to organize your thinking in rule-using tasks.

[X] **Rule Finding** is the last and most difficult of the four skills. It is the ability to improvise a solution to a problem when no other skill applies.

Rule finding presents something of a paradox to me as a teacher. On the one hand, I'd like it to be the single most important ability you carry away from this course. Rule finding is the product of real education; it is what remains after you forget everything you know. If you can apply the content of this course to the myriad new and unpredictable situations you'll encounter in life, then you'll have been successfully educated. On the other hand, an introductory course like this is not designed to make a chemist out of you. Its purpose is simply to give you an overview of the field. Thus, the teaching is aimed more at making you knowledgeable than at making you innovative.

As a practical matter, rule finding has no specific objectives; none is marked with an **[X]**. Nevertheless, in this study guide, which is imperfect despite my best efforts, you'll find places where I've failed to anticipate your needs. Whenever that happens, use your native intelligence to conquer the problem. You can do it.

What's stopping you, then? Not a thing. So go for it, and get all you can. (Honor your commitment.) Think of what it will be like when you see all those friends and relatives who told you how badly they did in chemistry. "I don't know what happened to them," you could say, "but *I* can do chemistry."

Chapter 1

Chemistry:
Its Nature and Impact on Society

Outline

I. Our Chemical Age: Opportunities and Responsibilities
   - A. Reasons to study chemistry
   - B. Impact of chemistry on everyone
     1. Benefits
     2. Side effects

II. The Nature of Science and Technology: The Search for Order
   - A. What is Science?
     1. Misconceptions about science
     2. What scientists do
        - a. Data and laws
        - b. Hypotheses and theories
   - B. What is Technology?
     1. Applied science
     2. Technology vs. science
   - C. The Methods of Science
     1. The scientific method
     2. How it is applied in real world

III. The Nature of Chemistry: The Classification of Matter and Energy
   - A. Matter and Energy
     1. Matter vs. energy
     2. Chemistry
   - B. Classifying Matter
     1. Levels of organization
        - a. Subatomic particles
        - b. Atoms
        - c. Molecules
     2. Physical State
        - a. Solids, liquids, and gases
        - b. Temperature and pressure
        - c. Changes of state
          - (1) Melting and freezing
          - (2) Boiling and condensation
     3. Composition
        - a. Heterogeneous matter
        - b. Homogeneous matter
          - (1) Pure substances
            - (a) Elements
            - (b) Compounds
          - (2) Solutions
   - C. Physical and Chemical Changes
     1. Physical changes

2. Chemical changes
    a. Reactants and products
    b. Chemical equations
D. Classifying Energy
    1. Kinetic energy
    2. Potential energy

IV. Measuring Matter and Energy: The Importance of Numbers and Units
    A. Measurement
    B. Metric System of Units
    C. Some Common Units for Measurements of Matter
        1. Length units
        2. Volume units
        3. Mass vs. weight units
    D. Some Common Units for Measurements of Energy and Temperature
        1. Energy units
        2. Temperature units
    E. Chemistry and the Future

Objectives

After you read and study the chapter [and the sections in brackets], you should be able to:

[C] 1. Distinguish science from technology and identify examples of each. [Section 1.2]

[C] 2. Identify examples of scientific data, scientific laws, scientific hypotheses, and scientific theories. [Section 1.2; Questions 1, 2, 3, 4]

[C] 3. Distinguish between or among:
    a. Matter and energy
    b. Solids, liquids, and gases
    c. Melting, boiling, condensing, and freezing
    d. Homogeneous and heterogeneous matter
    e. Pure substances, mixtures, elements, compounds, solutions
    f. Physical and chemical changes

    Identify examples of each. [Section 1.3; Questions 5, 6, 7]

[M] 4. Give an appropriate metric unit for measuring length, volume, mass, energy, and temperature. [Section 1.4]

[M] 5. Recall the commonly used metric prefixes and multiplier factors in Table 1.1. [Section 1.4]

Practice

Objective 1: Distinguish science from technology and identify examples of each.

Science is the systemized knowledge of nature gathered for its own sake.

Technology is the use or application of science for the creation of new products and processes.

Examples:

1. The invention of the atomic bomb. TECHNOLOGY--product created.
2. The discovery of atomic energy. SCIENCE--just knowledge.
3. Some substances, called semiconductors, conduct electricity only slightly. SCIENCE--just knowledge.
4. Computer chips are made out of semiconductors. TECHNOLOGY--useful product.

Drill:

1. Circle S for science or T for technology:
   a) Moistened baking powder will give off bubbles of gas. **S T**
   b) Lithium compounds can be used as medicines to treat patients with depression. **S T**
   c) Aluminum metal is recovered from its ores by the Hall process. **S T**
   d) A simple equation describes how gases expand when they are heated up. **S T**
   e) A tungsten filament, like those in light bulbs, will glow brightly when electricity is passed through it. **S T**
2. Which one of the following is NOT a product of technology?
   a) gasoline additives b) aspirin
   c) table salt d) Teflon
3. Is noticing that coal burns giving off large amounts of heat an example of science or technology?

Objective 2: Identify examples of scientific data, scientific laws, scientific hypotheses, and scientific theories.

Scientific **data** and **laws** are facts gathered from observation without concern for why they occur. Data are individual facts, and laws are generalizations of many facts that predict how nature will behave in the future.

Scientific **hypotheses** and **theories** are statements of possible laws of nature. They are interpretations or explanations for scientific data and laws. Hypotheses are initial, working conjectures that lack extensive testing or widespread acceptance. Theories are well-tested and widely accepted ideas.

Examples:

1. The sun came up in the east yesterday. DATA--an individual observation.
2. The sun always comes up in the east. LAW--prediction of future events.
3. The sun moves around the earth. HYPOTHESIS--explains the law in (2) but is not a widely-accepted view.
4. The earth moves around the sun. THEORY--widely-accepted and well-tested idea.

Drill:

1. Classify the following as Data, Law, Hypothesis, or Theory:

a)	The moon is made of green cheese.	**D**	**L**	**H**	**T**
b)	Matter is made of atoms and molecules.	**D**	**L**	**H**	**T**
c)	A sample of table salt contains sodium.	**D**	**L**	**H**	**T**
d)	What goes up must come down.	**D**	**L**	**H**	**T**
e)	Cigarettes cause cancer.	**D**	**L**	**H**	**T**
f)	The earth is flat.	**D**	**L**	**H**	**T**
g)	An earthquake measured 5.2 on the Richter scale.	**D**	**L**	**H**	**T**
h)	One can get AIDS from a sneeze.	**D**	**L**	**H**	**T**
i)	Toads cause warts.	**D**	**L**	**H**	**T**
j)	Germs cause disease.	**D**	**L**	**H**	**T**
k)	Compressed gas will expand.	**D**	**L**	**H**	**T**
l)	The Hope diamond is yellowish.	**D**	**L**	**H**	**T**

Objective 3a: Distinguish between matter and energy.

Matter is anything that has mass and occupies space. **Energy** is the capacity to act vigorously and to cause movement.

Examples:

1. Consider a spinning wheel. The wheel itself (whether still or moving) is MATTER; its spin is ENERGY.
2. Imagine a flame. Any soot, smoke, or vapor is MATTER; any light or heat is ENERGY.

Drill:

1. Classify as Matter or Energy:

a)	the "coldness" inside a refrigerator	**M**	**E**
b)	a jolt of electricity	**M**	**E**
c)	a gnat's eyelash	**M**	**E**

d)	the warmth of the sun	**M**	**E**
e)	a ton of bricks	**M**	**E**

Objective 3b: Distinguish among solids, liquids, and gases.

Your intuitive ideas about these states of matter are probably correct. **Solids** have a definite shape and a fixed volume no matter what container they occupy. **Liquids** also have a fixed volume, but they take the shape of their container. **Gases** take both the shape and volume of their container.

Examples:

1. Water is a LIQUID; steam is a GAS; and ice is a SOLID. All three are different states of the same substance.
2. Thick syrup and taffy are LIQUIDS; they don't have to be runny.
3. Dust particles in the air are SOLIDS; they don't have to massive.

Drill:

1. Tell whether the following are Gases, Liquids, or Solids:

a)	Jello	G	L	S
b)	Smoke	G	L	S
c)	Gasoline	G	L	S
d)	Smog	G	L	S
e)	Molten lava	G	L	S
f)	Helium	G	L	S

Objective 3c: Distinguish among melting, boiling, condensing, and freezing.

These are all changes in state. **Melting** is the change from solid to liquid; **freezing** is the opposite change from liquid to solid. Both occur at the same temperature. **Boiling** is the change from liquid to gas at a characteristic temperature called the boiling point (most liquids also tend to evaporate--change to gas at a lower temperature; this is not boiling); **condensing** is the change from gas to liquid.

Examples:

1. Dew forming in the morning is an example of CONDENSING (water vapor to liquid).
2. Candle wax dripping from around the burning wick is an example of MELTING (solid to liquid).
3. The sizzling of drippings from a steak on a barbeque is an example of BOILING (liquid to gas).
4. Icicles form by FREEZING (liquid to solid).

Drill:

1. Indicate whether the each of the following changes is **M**elting, Freezing, Boiling, or Condensing:

a.	ice forming on a lake	M	F	B	C
b.	fogging up a mirror	M	F	B	C
c.	bubbling oil for deep fat frying	M	F	B	C
d.	pat of butter becoming a yellow puddle	M	F	B	C

Objective 3d: Distinguish between homogeneous and heterogeneous matter.

Homogeneous matter appears to be the same through and through. The emphasis is on appearance rather than content.

Heterogeneous matter appears uneven and nonuniform. This depends on how close you look. Irregularities not apparent to the naked eye may become obvious under a microscope. But most of the time heterogeneous (like homogeneous) indicates how matter looks to the unaided eye.

Examples:

1. Fruit punch is HOMOGENEOUS because each visible drop of it looks identical to every other even though the punch may actually be a mixture of water, sugar, fruit juices, and colorings.
2. Wood is HETEROGENEOUS because the dark and light grain markings are visible irregularities.
3. Chunks of ice are HOMOGENEOUS. Even though they may be of random sizes and shapes, each chunk is the same through and through.
4. Beach sand is HETEROGENEOUS. Although it may seem uniform from afar, the individual grains of sand have visible color differences.

Drill:

1. Are the following homogeneous (O) or heterogeneous (X)?

a.	chocolate chip cookies	O	X
b.	powdered sugar	O	X
c.	beer	O	X
d.	house paint	O	X
e.	the comic section of the Sunday paper	O	X
f.	grapefruit	O	X
g.	aspirin	O	X
h.	asphalt	O	X
i.	sulfur	O	X

Objective 3e: Distinguish among pure substances, mixtures, elements, compounds, and solutions.

A **pure substance** has a fixed composition, contains one kind of matter through and through, and cannot be separated into other substances by physical means. Other connotations of the word pure are irrelevant in a chemical context; a pure substance is not necessarily wholesome, virtuous, or of a single origin.

A **mixture** is the combination of two or more different substances in the same container. These substances can be separated by physical means like evaporation or filtering, and their composition is arbitrary.

An **element** is a pure substance that cannot be broken down into simpler substances by chemical or physical means. It is tabulated on the list of elements inside the back cover of the textbook.

A **compound** is a pure substance consisting of two or more elements that can only be separated by chemical means. Its composition is fixed. It is not a mixture.

A **solution** is a homogeneous mixture. Often found in the liquid state (though not limited to liquids), a solution only appears to be the same throughout. Certain physical means like evaporation or distillation can always show it to be made of two or more different substances.

Note that matter is classified (see Figure 1.5 in the textbook) by its composition (fixed or arbitrary) and by its ability or inability to be separated (chemically or physically). Both these criteria require you to know some specifics about the matter beyond its mere identity.

Examples:

1. Baking soda is a common PURE SUBSTANCE. A box of it contains only one kind of matter that can't be filtered or separated except by reacting it chemically (as with vinegar or stomach acid, for example). It is not on the list of elements; therefore, it is a COMPOUND.
2. Aluminum foil is also a PURE SUBSTANCE. But its one kind of matter cannot be separated even chemically. It's on the element list, so it must be an ELEMENT.
3. Brass, like all metal alloys, is a MIXTURE of at least two metals with variable composition. Because it is homogeneous, it must be a SOLUTION.
4. Vinegar and salad oil is a MIXTURE. Since it tends to separate automatically into two layers, and since it never gets fully homogeneous even when it is well-shaken, it is not a solution.

Drill:

1. Classify the following pure substances as Elements or Compounds:

a. 200 proof alcohol E C
b. gypsum E C
c. platinum E C
d. radium E C
e. silica E C
f. silicone E C
g. silicon E C

2. Tell whether the following are **P**ure substances or **M**ixtures:
   a. carbon monoxide P M
   b. honey P M
   c. mayonnaise P M
   d. hospital oxygen P M
   e. jello P M
   f. air P M
   g. distilled water P M

3. Identify the mixtures that are Solutions and those that are **N**ot Solutions
   a. Thousand Island salad dressing S N
   b. Raisin Bran S N
   c. pancake syrup S N
   d. wine S N
   e. cooked oatmeal S N
   f. stainless steel S N
   g. gasoline S N

Objective 3f: Distinguish between physical and chemical changes.

A **physical change** is one that occurs without changing the identity of the substance.

A **chemical change** is the reaction between two or more substances that produces a new substance.

Examples:

1. Iron rusting is a CHEMICAL CHANGE. The brown rust is clearly different from the iron metal.
2. Water boiling is a PHYSICAL CHANGE because steam is not a new substance but is simply gaseous water. (All changes of state are physical changes.)
3. Sugar dissolving in water is a PHYSICAL CHANGE. The sugar does not become a new substance, rather it simply mixes with the water. Nearly all dissolvings are physical changes.
4. The spoiling of food is a CHEMICAL CHANGE. New substances, which you can taste, are formed.

Drill:

1. Identify each of the following as a **P**hysical or a **C**hemical change.
   a. Soda pop fizzes P C
   b. A grinding wheel produces sparks P C

c.	Banana slices turn dark	P	C
d.	Dry ice evaporates	P	C
e.	Bread browns in the toaster	P	C
f.	A bathtub ring forms in soapy water	P	C
g.	Milk goes sour	P	C
h.	The wind blows	P	C
i.	Polluted rain water becomes acidic	P	C
j.	Flashlight batteries discharge	P	C

Objective 4: Give an appropriate metric unit for measuring length, volume, mass, energy, and temperature.

Drill:

Cut out the flash cards, and memorize the material on them.

Objective 5: Recall the commonly used metric prefixes and multiplier factors in Table 1.1.

Drill:

Cut out the flash cards, and memorize the material on them.

Self-Test

1. In the metric system, one measures energy in
   a) grams b) calories c) watts d) meters
2. A kiloton bomb is equivalent to how many tons?
3. Melting is a physical change. True or false?
4. Are teflon-coated frying pans a product of science or technology?
5. What goes up must come down. This is a scientific
   a) data b) law c) theory d) hypothesis
6. Is homogenized milk really homogeneous?
7. A millimeter is __________ (longer, shorter) than a centimeter.
8. Table salt is a
   a) compound b) element c) mixture d) solution
9. Which is not a proper metric unit of volume?
   a) m^3 b) cc c) mL d) mm
10. Matter is made up of atoms and molecules. Is this
    a) data b) law c) theory d) hypothesis
11. Which is science rather than technology, electricity or electronics?
12. When snowflakes form in a cloud, this is
    a) melting b) freezing c) boiling d) condensing
13. True or false: a super-thick milkshake is a liquid not a solid.
14. Is a bolt of lightening matter or energy?

Answers

Objective 1:

1. a) S; b) T; c) T; d) S; e) S
2. c)
3. science

Objective 2:

1. a) H, b) T, c) D, d) L, e) T, f) H, g) D, h) H, i) H, j) T, k) L, l) D

Objective 3a:

1. a) E; b) E; c) M; d) E; e) M

Objective 3b:

1. a) S; b) S; c) L; d) G; e) L f) G

Objective 3c:

1. a) F; b) C; c) B; d) M

Flash Cards for Objectives 4 and 5

Length	Volume
Mass	Energy
Temperature	micro-
milli-	centi-
kilo-	mega-

cubic meters (m^3) or liters (L)	meters (m)
joules (J) or calories (cal)	grams (g)
one-millionth	degrees Celsius (^oC)
one-hundredth	one-thousandth
a million times	a thousand times

Objective 3d:

1. a) X; b) O; c) O; d) O; e) X; f) X; g) O; h) X; i) O

Objective 3e:

1. a) C; b) C; c) E; d) E; e) C; f) C; g) E
2. a) P; b) M; c) M; d) P; e) M; f) M; g) P
3. a) N; b) N; c) S; d) S; e) N; f) S; g) S

Objective 3f:

1. a) P; b) P; c) C; d) P; e) C; f) C; g) C; h) P; i) C; j) C

Self-Test:

1) b; 2) 1000; 3) true; 4) technology; 5) theory; 6) yes; 7) shorter; 8) compound; 9) d; 10) theory; 11) electricity; 12) freezing; 13) true; 14) energy

Evaluation:

If you missed more than one question on the self-test in any of the following groups, you need to review the section indicated:

Question Groups	Section
4, 5, 10, 11	1.2
3, 6, 8, 12, 13, 14	1.3
1, 2, 7, 9	1.4

Chapter 2

Atoms as Building Blocks of Matter: Atomic Structure and the Periodic Table

Outline

I. Early Greek Models of the Atom: Pure Imagination
   A. What Happens If We Keep Subdividing Matter?
   B. "You Get Atoms," said Leucippus and Democritus
   C. "No, You Don't," said Plato and Aristotle
II. Dalton's Atomic Theory: A Model Based on Experiments
   A. Lavoisier: The Law of Conservation of Mass
   B. Proust: The Law of Constant Composition
   C. Dalton's Atomic Theory
   D. Dalton: The Law of Multiple Proportions
III. Atomic Particles: What's Inside is Important
   A. Discovery of the Electron
      1. Cathode ray tube experiments
      2. J. J. Thomson's conclusions
   B. Discovery of the Proton
      1. Modified cathode ray tube experiments
      2. Eugen Goldstein's conclusions
   C. Thomson Model of the Atom
   D. Discovery of the Atomic Nucleus: The Rutherford Model
      1. Radioactivity
         a. Alpha particles
         b. Beta particles
         c. Gamma rays
      2. Ernest Marsden's experiment
      3. Ernest Rutherford's conclusions
   E. Discovery of the Neutron
      1. Walther Bothe's experiments
      2. James Chadwick's conclusions
   F. Subatomic particles: A Summary
IV. Models of Electronic Structure: How Are the Electrons in Atoms Arranged?
   A. Electronic Structure
   B. Line Spectra: Emission of Light by Elements
      1. White light; continuous spectrum
      2. Line spectrum; discontinuous
   C. The Bohr Model
      1. Major energy levels
      2. Ground states compared to excited states
      3. Planetary model and its flaws
   D. The Wave Mechanical Model of Electronic Structure
      1. Energy sublevels
      2. Mathematical probability compared to exact electron locations
   E. Atomic Size

V. Atomic Number, Mass Number, Isotopes, and Atomic Mass: Some Important Characteristics of Atoms
  A. Atomic Number: A Way to Distinguish One Element from Another
    1. Numbers of protons
    2. Elements defined in terms of atomic number
  B. Neutrons, Isotopes, and Mass Numbers
    1. Atoms with same number of protons but different numbers of neutrons
    2. Mass numbers
    3. Atomic symbols
  C. Atomic Mass or Weight of an Element
    1. Relative masses
    2. Atomic mass units
    3. Average masses of naturally occurring isotopes
VI. The Periodic Table: A Chemical Masterpiece
  A. Development of the Periodic Table
  B. The Modern Periodic Table
    1. Periods
    2. Groups or families
    3. Main group elements
      a. Alkali metals
      b. Alkaline earths
      c. Halogens
      d. Noble gases
  C. Major Types of Elements
    1. Metals
      a. Transition metals
      b. Inner transition metals
    2. Nonmetals
      a. Noble gases
      b. Others
    3. Metalloids
  D. The Periodic Table and Electronic Structure
    1. Valence electrons
    2. Electron dot structure

Objectives

After you read and study the chapter [and the sections in brackets], you should be able to:

[M] 1. Describe the following models of the atom:
  a) Democritus [Section 2.1; Question 1]
  b) Dalton [Section 2.2; Question 2]
  c) Thomson [Section 2.3; Questions 3, 4, 5]
  d) Rutherford [Section 2.3; Questions 5, 6]
  e) Bohr [Section 2.4; Questions 13, 14, 15, 16]
  f) Wave Mechanical [Section 2.4; Questions 15, 16]

[M] 2. Associate the names and the chemical symbols of the 15 elements listed in Table 2.4, page 45. [Section 2.6]

[R] 3. Determine, for any element, all of the following quantities, given one from each group:
- o Atomic number, chemical symbol, number of protons
- o Mass number, number of neutrons
- o Electrical charge, number of electrons

[Section 2.5; Questions 7, 8, 9, 10, 11, 12]

[C] 4. Cite or identify examples of
- a) metals, nonmetals, and metalloids. [Section 2.6, Question 18]
- b) alkali metals, alkaline earths, halogens, noble gases, transition elements, and inner transition elements. [Section 2.6, Question 18]
- c) other elements that are chemically similar to a specified element. [Section 2.6; Questions 17, 19, 20]

Practice

Objective 1: Describe the following models of the atom: a) Democritus; b) Dalton; c) Thomson; d) Rutherford; e) Bohr; f) Wave Mechanical

Drill:

Cut out the flash cards and memorize the material on them.

Objective 2: Associate the names and the chemical symbols of the 15 elements listed in Table 2.4, page 45.

Drill:

Cut out the flash cards and memorize the material on them.

Flash Cards for Objectives 1 and 2

The Democritus Model of the Atom	The Dalton Model of the Atom
The Thomson Model of the Atom	The Rutherford Model of the Atom
The Bohr Model of the Atom	The Wave-Mechanical Model of the Atom
H	C
O	N

The atom is the basic building block of an element, and it cannot be created, destroyed, subdivided, or converted into atoms of another element.

The atom is the smallest particle of matter, and it cannot be subdivided

The atom is a tiny positively-charged nucleus surrounded by empty space through which electrons travel.

The atom is a positively-charged sphere with electrons imbedded in it like seeds in a watermelon.

The atom is a complicated entity that can be described only by equations involving higher matihematics.

The atom is a nucleus surrounded by electrons that reside in certain allowed energy levels only.

carbon

hydrogen

nitrogen

oxygen

P	S
F	Na
Al	Si
Cl	K
Ca	Fe

sulfur	phosphorus
sodium	fluorine
silicon	aluminum
potassium	chlorine
iron	calcium

Objective 3: Determine, for any element, all of the following quantities, given one from each group:

- o Atomic number, symbol, number of protons
- o Mass number, number of neutrons
- o Electrical charge, number of electrons

Rule:

1. The atomic number is the same as the number of protons. The atomic number and chemical symbol for each element are listed on the periodic table.
2. The mass number is the number of protons plus the number of neutrons. The number of neutrons is the mass number minus the number of protons.
3. The electrical charge is the number of protons minus the number of electrons. The number of electrons is number of protons minus the charge.

Procedure: Find the missing quantities in each group, working in the order listed above.

Examples:

1. Fill in the following table:

	Atomic Number	Symbol	$\#p^+$	Mass Number	$\#n^0$	Charge	$\#e^-$
a)	57	______	______	139	______	0	______
b)	______	______	34	______	45	______	36
c)	______	Cl	______	35	______	-1	______

a) Step 1: #p+ = at.no. = 57; from table, element 57 is La. Step 2: $\#n^0$ = mass no. - $\#p^+$ = 139 - 57 = 82. Step 3: $\#e^-$ = $\#p^+$ - chrg = 57 - 0 = 57

b) Step 1: at.no. = $\#p^+$ = 34; from table, element 34 is Se. Step 2: mass no. = $\#p^+$ + $\#n^0$ = 34 + 45 = 79. Step 3: chrg = $\#p^+$ - $\#e^-$ = 34 - 36 = -2

c) Step 1: from table, Cl is element 17; $\#p^+$ = at.no. = 17. Step 2: $\#n^0$ = mass no. - $\#p^+$ = 35 - 17 = 18. Step 3: $\#e^-$ = $\#p^+$ - chrg = 17 - (-1) = 18

Thus:

	Atomic Number	Symbol	$\#p^+$	Mass Number	$\#n^0$	Charge	$\#e^-$
a)	57	La	57	139	82	0	57
b)	34	Se	34	79	45	-2	36
c)	17	Cl	17	35	18	-1	18

Drill:

1. Fill in the following table:

	Atomic Number	Symbol	#p^+	Mass Number	#n^0	Charge	#e^-
a)	______	As	______	______	42	-3	______
b)	64	______	______	155	______	______	61
c)	______	______	12	25	______	+2	______
d)	______	I	______	______	74	______	54
e)	48	______	______	115	______	______	48
f)	58	______	______	______	82	+4	______
g)	______	______	16	32	______	-2	______

Objective 4a: Cite or identify examples of metals, nonmetals, and metalloids.

The **metalloids**, being borderline cases between metals and nonmetals, are pictured in gray on the periodic table inside the front cover of the textbook.

Metals are all pictured in solid orange to the left and below the metalloids on the periodic table.

Nonmetals are either light orange or unshaded to the right and above the metalloids on the periodic chart.

Hydrogen (top center) is a nonmetal.

Examples:

1. Ga: Atomic number 31; a METAL in solid orange.
2. Ge: Atomic number 32; a METALLOID in gray.
3. Kr: Atomic number 36; a NONMETAL unshaded.

Drill:

1. Identify the following elements as Metals, Nonmetals, or MetallOids:
   - a) Sb M N O
   - b) Tm M N O
   - c) W M N O
   - d) C M N O
   - e) Br M N O
   - f) At M N O
   - g) He M N O
2. Give the chemical symbol of
   - a) a metalloid in Period 3
   - b) a nonmetal in Period 6
   - c) a metal in Period 2

Objective 4b: Cite or identify examples of alkali metals, alkaline earths, halogens, noble gases, transition elements, and inner transition elements.

Alkali metals are all elements in Group 1/IA. **Alkaline earths** are all elements in Group 2/IIA. **Halogens** are all elements in Group 17/VIIA. **Noble gases** are all elements in Group 18/VIIIA. **Transition elements** are all those in the short B columns (3 through 12 inclusive). **Inner transition elements** are all those in bottom two rows labelled Lanthanides (rare earth elements) and Actinides.

Some elements, namely those in Groups 13/IIIA to 16/VIA, do not belong to any of these categories.

Examples:

1. Zn is a TRANSITION ELEMENT; I is a HALOGEN; U is an INNER TRANSITION ELEMENT; Rb is an ALKALI METAL; He is a NOBLE GAS; Be is an ALKALINE EARTH. Pb is NONE OF THESE.

Drill:

1. Tell whether the following elements are Alkali metals, Alkaline Earths, Halogens, Noble gases, Transition elements, Inner transition elements, or Other:

a)	V	A	E	H	N	T	I	O
b)	As	A	E	H	N	T	I	O
c)	Ar	A	E	H	N	T	I	O
d)	Dy	A	E	H	N	T	I	O
e)	Ca	A	E	H	N	T	I	O
f)	Ra	A	E	H	N	T	I	O
g)	Li	A	E	H	N	T	I	O
h)	H	A	E	H	N	T	I	O
i)	Hg	A	E	H	N	T	I	O

2. Give the symbol of
   a) a transition element in Group 7/VIIB
   b) an alkaline earth in Period 4
   c) an inner transition element starting with C
   d) a metalloid halogen

Objective 4c: Cite or identify other elements that are chemically similar to a specified element.

Each group (vertical column) on the periodic table contains elements that are chemically similar to one another. Thus, the elements that react like some given element are those in the same column.

(Elements in different columns may be similar also, but we'll ignore that possibility here.)

Examples:

1. Give an example of an element that is chemically similar to iron. Fe is in Group 8; thus, Ru or Os would be most similar.
2. Which would be most like table salt or sodium chloride, potassium chloride or magnesium chloride? Potassium chloride because both Na and K are in Group 1/IA.

Drill:

1. Match the following elements with that element in the right hand column which is most chemically similar.

a)	Ag	Na
b)	Si	Mg
c)	Cs	Cu
d)	O	Al
e)	He	C
f)	Sr	S
g)	Tl	Ne

Self-Test

---

1. The first model of the atom to deal with electrons was that of
   a) Rutherford b) Thomson c) Bohr d) Dalton
2. What are the symbols for calcium and potassium?
3. How many neutrons in iron-56?
4. Name a metalloid in Group 14/IVA.
5. Give names of Na and S.
6. Which is not part of the Bohr model of the atom?
   a) electrons reside in allowed energy levels
   b) the atom has a nucleus
   c) electrons are imbedded in the nucleus
   d) the number of protons gives the atom its identity
7. What was Democritus' contribution to the atomic theory?
8. Which list contains an alkaline earth and a transition element?
   a) V, Na b) Sr, Pb c) Ba, Tl d) Re, Be
9. What charge would an atom have if it contained 14 neutrons, 12 protons, and 10 electrons?
10. What is the mass number and atomic number of the atom in Question 9?
11. Which element is named incorrectly?
    a) Si, silicone b) C, carbon
    c) Cl, chlorine d) Al, aluminum
12. Which of the following would be most similar to phosphorus?
    a) B b) N c) S d) O

Answers

Objective 3:

1.a)	33	As	33	75	42	-3	36
b)	64	Gd	64	155	91	+3	61
c)	12	Mg	12	25	13	+2	10
d)	53	I	53	127	74	-1	54
e)	48	Cd	48	115	67	0	48
f)	58	Ce	58	140	82	+4	54
g)	16	S	16	32	16	-2	18

Objective 4a:

1. a) O; b) M; c) M; d) N; e) N; f) N; g) N
2. a) Si; b) Rn; c) Li or Be

Objective 4b:

1. a) T; b) O; c) N; d) I; e) E; f) E; g) A; h) O; i) T
2. a) Mn or Tc or Re; b) Ca; c) Ce or Cm or Cf; d) At

Objective 4c:

1. a) Cu; b) C; c) Na; d) S; e) Ne; f) Mg; g) Al

Self-Test:

1) b; 2) Ca, K; 3) 30; 4) Si or Ge; 5) sodium and sulfur; 6) c; 7) he was the first to think of it; 8) d; 9) +2; 10) 26; 11) a (silicon); 12) b

Evaluation:

If you missed more than one question on the self-test in any of the following groups, you need to review the section indicated:

Question Groups	Section
1, 6, 7	2.1 to 2.4
3, 9, 10	2.5
2, 4, 5, 8, 11, 12	2.6

Chapter 3

Nuclear Chemistry: The Core of Matter

Outline

I. The Nature of Radioactivity: Spontaneous Emissions From Unstable Nuclei
    A. Discovery of Radioactivity
        1. Henri Becquerel's uranium rock
        2. Madame Curie's radium and polonium
    B. Major Types of Radioactivity
        1. Alpha particles
        2. Beta particles
        3. Gamma rays

II. Natural Radioactivity and Nuclear Reactions: Representing Nuclear Changes
    A. Nuclear reactions and nuclear equations
    B. Writing nuclear equations
        1. Alpha emissions
        2. Beta emissions
        3. Gamma emissions
        4. Identifying unknown isotopes

III. Half-Life: How Long Does Radioactivity Last?
    A. Rates of radioactive emission
    B. Half-life
    C. Decay curves

IV. Artificial Radioactivity: Making New Isotopes and Elements
    A. Producing Synthetic Isotopes
        1. Radioactivity not found in nature
        2. Positrons
    B. Producing New Elements
        1. Cyclotrons and linear accelerators
        2. Elements beyond uranium

V. Harmful Effects of Radioactivity: How Much Is Too Much?
    A. Effects of Radioactivity
        1. Outside vs. inside the body
        2. Rems
    B. Background and Human-Induced Radiation
        1. Natural or Background radiation
            a. Cosmic rays
            b. Natural radioisotopes in the surroundings
            c. Natural radioisotopes ingested into body
        2. Radiation from human activities
            a. Medical and dental X rays
            b. Nuclear power plants
    C. How Much Radiation is Harmful?
        1. Estimates by government agencies
        2. Effects of smoking

VI. Useful Applications of Radioisotopes: Tracing Chemicals and Saving Lives
   A. Determining the Age of Objects
      1. Radiocarbon dating
      2. Potassium-argon dating
   B. Uses of Radioactivity in Industry, Agriculture, and Research
      1. Tracers
      2. Irradiated seeds
      3. Pest control
   C. Uses of Radioisotopes in Medicine
      1. Powering pacemakers
      2. Diagnosis
         a. Tracing bloodflow
         b. Measuring function of thyroid and other organs
         c. Detecting brain tumors
      3. Cancer treatment

VII. Nuclear Fission and Nuclear Fusion: Splitting and Combining Nuclei
   A. Splitting Heavy Nuclei: Nuclear Fission
      1. Early evidence of fission
      2. Energy from fission
         a. $E = mc^2$
         b. Chain reactions
         c. Atomic bombs
   B. Joining Light Nuclei: Nuclear Fusion
      1. Fusion in the universe
      2. Hydrogen bombs
   C. Nuclear War: The Ultimate Pollution

Objectives

After you read and study the chapter [and the sections in brackets], you should be able to:

[C] 1. Classify the three major types of natural radiation by the characteristics of each. [Section 3.1; Questions 2, 3]

[R] 2. Write balanced equations to represent nuclear reactions. [Section 3.2; Question 5]

[R] 3. Use half-life values to calculate the length of time radioactivity lasts for a particular isotope. [Sections 3.3, 3.4; Questions 1, 8]

[C] 4. Describe harmful effects and useful applications of radioactivity. [Sections 3.5, 3.6; Questions 4, 6, 7, 11]

[C] 5. Explain what fission and fusion are, and tell how they may be used. [Section 3.7; Questions 9, 10]

Practice

Objective 1: Classify the three major types of natural radiation by the characteristics of each.

The three major types of radiation emitted by natural radioactive materials are called alpha, beta, and gamma emissions.

Alpha emissions consist of two protons and two neutrons. They are the same as helium-4 nuclei and have a net electrical charge of 2+ because of the two protons. Compared to beta and gamma emissions, alpha emissions move at the slowest speed and are the least penetrating, but cause the most damage inside the body.

Beta emissions are electrons coming from the nucleus. They have a 1- electrical charge. Compared with alpha and gamma emissions, beta emissions have an intermediate speed, penetration, and ability to damage the body.

Gamma emissions are high-energy electromagnetic radiation. Compared with alpha and beta emissions, gamma emissions travel the fastest, are the most penetrating, and do the most damage from outside the body (due to their penetration) but the least damage inside the body.

Examples:

1. Thin aluminum foil can stop ALPHA emissions. An aluminum plate 0.3 cm (one-eighth inch) thick can stop BETA emissions, but not GAMMA emissions.
2. A tiny amount of plutonium-239, an ALPHA emitter, is very dangerous when ingested in the body, but you can safely hold the same amount of it in your hand (with gloves on) for a short time.
3. When passing through a magnetic field, ALPHA and BETA emissions (which have electrical charges) are deflected a bit from their original path; GAMMA emissions (which have no electrical charge) are not affected.

Drill:

1. Isotopes that have to be surrounded by heavy lead containers emit what kind of radiation?
2. Rutherford and Marsden, in their gold foil experiment (Section 2.3), used a radioactive sample whose emissions were repelled by nuclei. What type of emissions were these?
3. If you had to ingest a radioactive material for a medical diagnosis, which type of emitter would be the most dangerous to use?

Objective 2: Write balanced equations to represent nuclear reactions.

Rule:

1. A balanced equation shows the reacting material(s) on the left of an arrow and the product(s) on the right.
2. Each isotope is represented by its appropriate symbol preceded by its atomic number as a subscript and its mass number as a superscript.
3. The symbols for the radioactive emissions are:

   alpha: ${}^{4}_{2}He$; beta: ${}^{0}_{-1}e$; and gamma: γ (optional)

   Other symbols are:

   proton: ${}^{1}_{1}p$; neutron: ${}^{1}_{0}n$; positron: ${}^{0}_{+1}e$

4. The sums of the atomic numbers on each side of the arrow must be equal, and the sums of the mass numbers on each side of the arrow must be equal.

Procedure:

1. Find the mass number of the missing material by determining what integer will make the sums of the mass numbers equal.
2. Find the atomic number of the missing material by determining what integer will make the sums of the atomic numbers equal.
3. Use the periodic chart or #3 above to determine the symbol that corresponds to the atomic number.

Examples:

1. Complete: ${}^{90}_{38}Sr$ --> ${}^{90}_{39}Y$ + ______

   Step 1: the mass numbers are already equal on both sides; therefore, the missing mass number must be 0. Step 2: the missing atomic number must be -1. Step 3: the missing substance must be an electron, ${}^{0}_{-1}e$.

2. What isotope decays to yield the following products?

   ______ --> ${}^{231}_{90}Th$ + ${}^{4}_{2}He$ + γ

   Step 1: The mass number is 231 + 4 = 235. Step 2: The atomic number is 90 + 2 = 92. (Note that the gamma ray did not affect either sum.) Step 3: Element 92 is uranium; therfore, the isotope is ${}^{235}_{92}U$.

Drill:

1. Complete the following nuclear equations by writing the symbol for the missing component:

a) $^{137}_{55}Cs$ --> $^{0}_{-1}e$ + ________

b) $^{2}_{1}H$ + $^{3}_{1}H$ --> $^{1}_{0}n$ + ________

c) $^{95}_{40}Zr$ --> $^{95}_{41}Nb$ + γ + ________

d) ________ + $^{4}_{2}He$ --> $^{17}_{8}O$ + $^{1}_{1}P$

Objective 3: Use half-life values to calculate the length of time radioactivity lasts for a particular isotope.

Rule: For each half-life, a period of time characteristic of each radioactive isotope, one half of the remaining sample undergoes radioactive decay.

Procedure: Cut the remaining amount of material in half for every half-life of time that passes.

Examples:

1. Strontium-90 has a half-life of 28 years. What remains of a 96 g sample after 112 years?

0 yrs		28 yrs		56 yrs		84 yrs		112 yrs
96g	-->	48 g	-->	24 g	-->	12g	-->	**6 g**

2. The half-life of iodine-131 is 8 days. How many days have passed if a sample originally weighing 0.144 g has decayed to 0.009 g?

0.144 g	0.072 g	0.036 g	0.018 g	0.009 g
0 days	8 days	16 days	24 days	**32 days**

Drill:

1. How much will remain of a 7.2 mg sample of hydrogen-3 (half-life 12.5 years) after 50 years?
2. What will remain of a 0.1 g sample of carbon-14 after 11,460 years? Its half-life is 5,730 years.
3. The half-life of lead-214 is 19.7 minutes. How much time has passed if a 10.0 mg sample has decayed to 2.5 mg?
4. A 0.64 g sample of californium-244 decays to 0.16 g in 90 minutes. What is its half-life?

Objective 4: Describe harmful effects and useful applications of radioactivity.

Radiation is harmful to cells, particularly to their genetic material (DNA); thus, radiation can damage virtually all living things. Cells that reproduce rapidly tend to be the most vulnerable. In humans, this includes the bone marrow, spleen, gastrointestinal tract, reproductive organs, and lymph glands. Developing embryos and fetuses are also vulnerable.

Damage also occurs from nuclear weapons that use fission or fusion (see Objective 5 below).

Radiation is useful for determining the age of objects, and in tracing materials in pollution detection, scientific research, agriculture, and industry. Radioisotopes also are used for medical diagnosis and treatment of diseases such as cancer. Fission and fusion are ways to generate energy for many uses (see Objective 5).

Examples:

1. Gamma emitters (which can penetrate the skin) such as cobalt-60 and cesium-137 have been used to treat certain types of cancer.
2. People exposed to high doses of radiation typically have low blood counts, gastrointestinal distress, and (if they were pregnant during exposure) increased risk of babies with birth defects.
3. The use of radioactive materials in trace amounts has helped scientists discover many details about reactions that occur in the body.

Drill:

1. Why is carbon-14, which has a half-life of 5,730 years, not useful in estimating the age of objects that are about 100,000 years old?
2. Radiation is being used to sterilize insect pests and consumer products, including food. What kind of emitters are the most likely to be used for these purposes? Why?
3. Why are cancer cells often more vulnerable to radiation than many types of normal cells?
4. The Environmental Protection Agency has set what amount of radiation as the average amount allowed for a person per year from artificial, nonmedical sources?
   a) 500 rem b) 0.17 rem c) 0.007 rem d) 2.3 rem

Objective 5: Explain what fission and fusion are, and tell how they may be used.

Fission is the splitting of large nuclei into smaller ones with the release of large amounts of energy. The continuous fission of a sample is called a **chain reaction**. The amount of fissionable material necessary to sustain a chain reaction is called a **critical mass**.

Fusion is the joining of small nuclei into a larger nucleus with the release of large amounts of energy.

Both fission and fusion have been used to make bombs. Fission is being used to generate energy in nuclear power plants; controllable fusion energy is still in the development stage (see Section 11.4 for more details about this).

Examples:

1. Uranium-233, uranium-235, plutionium 239, and californium-242 are isotopes that undergo fission under appropriate conditions.
2. Hydrogen-2 (deuterium) amd hydrogen-3 (tritium) are the isotopes most commonly used for fusion.

Drill:

1. The bombs dropped on Nagasaki and Hiroshima during World War II were ________ (fission or fusion) bombs that used ________ as the fuel.
2. Which produces the most energy per gram of fuel used, fission or fusion?
3. Explain how the relationship between energy and mass helps account for the large amounts of energy released in fission and fusion.
4. What are the atomic numbers of the most stable nuclei, the ones with the lowest nuclear energy?
   a) 1-10 b) 20-30 c) 50-60 d) 100+

Self-Test

---

1. If a radioisotope has a half-life of 15 minutes, a 32 g sample will contain how much of the original isotope after one hour?
   a) 1 g b) 2 g c) 4 g d) 0.5 g
2. A radioisotope taken internally for medical diagnosis should have a ________ (short or long) half-life and be a ________ emitter.
3. The least penetrating type of emission from a natural radioactive material is ________
4. Certain isotopes much heavier than iron can release large amounts of energy by nuclear ________ (fission or fusion).
5. A type of cell particularly vulnerable to radiation damage is
   a) brain cells b) bone marrow cells
   c) liver cells d) muscle cells
6. What type of emission causes the atomic number of a nucleus to increase?
   a) alpha b) beta c) gamma d) neutron

7. A negatively-charged type of radioactive emission is
a) alpha b) beta c) gamma d) neutron
8. Radiation may be used for
a) tracing leaks in pipes b) treating cancer
c) sterilizing instruments d) all of these
9. If californium-242 emits an alpha particle, it will become
a) einsteinium-242 b) californium-238
c) curium-242 d) berkelium-241
10. How many half-lives must a radioisotope go through until there is 25% of the original isotope remaining?
a) 4 b) 3 c) 2 d) 1
11. Hydrogen-2 is an isotope used for
a) treating cancer b) dating old rocks
c) fusion d) fission
12. Fill in the missing component of the following equation:

$${}^{14}_{7}N + {}^{1}_{0}n \longrightarrow {}^{1}_{1}H + \_\_\_\_\_\_\_\_\_\_$$

13. Current nuclear power plants produce energy by nuclear ___________ (fission or fusion).
14. Background radiation from natural sources for most Americans is about
a) 130 millirems b) 5 rems
c) 220 rems d) 10 millirems

Answers

Objective 1:
1. gamma (most penetrating)
2. alpha (positively charged)
3. alpha (most damaging inside the body)

Objective 2:

1. ${}^{137}_{56}Ba$ 2. ${}^{4}_{2}He$ 3. ${}^{0}_{-1}e$ 4. ${}^{14}_{7}N$

Objective 3:
1. 0.45 g
2. 0.025 g
3. 39.4 min
4. 45 min

Objective 4:
1. Because of its half-life, hardly any carbon-14 would be left in an object after 100,000 years.
2. Gamma emitters, because they are more penetrating.
3. Cells that grow and reproduce rapidly--like cancer cells--are the most vulnerable to radiation damage.
4. b)

Objective 5:

1. Fission; uranium-235 and plutonium-239
2. Fusion
3. Mass and energy are two aspects of the same thing, and mass is converted into energy in fission and fusion. Since $E = mc^2$ and c^2 is a very large number, a small amount of mass produces a huge amount of energy.
4. c)

Self-Test:

1) b; 2) short, gamma; 3) alpha; 4) fission; 5) b; 6 b;

7) b; 8) d; 9) d; 10) c; 11) c; 12) $^{14}_{6}C$; 13) fission; 14) a

Evaluation:

If you missed more than one question on the self-test in any of the following groups, you need to review the section indicated:

Question Groups	Section
2, 3, 7	3.1
6, 9, 12	3.2
1, 2, 10	3.3
5, 8, 14	3.5, 3.6
4, 11, 13	3.7

Chapter 4

Chemical Bonding: Holding Matter Together

Outline

I. Chemical Bonds and the Octet Rule: Why Do Compounds Form?
   A. The Minimum Potential Energy Principle
      1. Lowest state of potential energy
      2. Electrostatic forces
         a. Attraction
         b. Repulsion
      3. Possibility of achieving minimum potential energy for certain combinations of atoms
         a. Forming ionic bonds
         b. Forming covalent bonds
         c. No possibility available
   B. The Octet Rule: Two and Eight as Magic Numbers
      1. Unreactive noble gases with stable electron configurations
         a. Two valence electrons (He)
         b. Eight valence electrons (Ne, Ar, Kr, Xe, Rn)
      2. Main group elements react to achieve stable electron configuration of nearest noble gas
   C. Bonding and the Octet Rule
      1. Ionic bonds: electrons transferred
      2. Covalent bonds: electrons shared
      3. Most bonds have both ionic and covalent character
   D. Predicting Principal Bond Type
      1. Metal-nonmetal: ionic
      2. Nonmetal-nonmetal: covalent

II. Metallic Bonding
   A. Properties of Metals and Nonmetals
      1. Physical properties
         a. Metals
            (1) Shiny
            (2) Workable (malleable and ductile)
            (3) Conductors (heat and electricity)
         b. Nonmetals
            (1) Dull
            (2) Brittle
            (3) Nonconductors
      2. Chemical properties
         a. Metals
            (1) Combine with nonmetals
            (2) Do not combine with each other
         b. Nonmetals
            (1) Combine with metals
            (2) Combine with each other

3. Trend of metallic character
    a. Increases toward left of a row
    b. Increases toward bottom of a column

B. Metallic Bonds and Metal Properties
    1. Metal crystal lattice
    2. Positive metal ions in a sea of electrons

III. Ionic Bonds: Giving and Taking Electrons

A. The Ionic Bond: Electron Transfer
    1. Attractions of positive and negative ions
    2. Formula units
    3. Ionic crystal lattice
    4. Ionic compounds
        a. Solids at room temperature with high melting points
        b. Good conductors of electricity in molten state

B. Writing Formulas and Formula Units for Simple Ionic Compounds
    1. Rule of electrical neutrality
    2. Crisscross rule

C. Naming Simple Two-Element Ionic Compounds
    1. Metal name - nonmetal name stem + "ide"
    2. Roman numeral convention

D. Formulas and Names of Ionic Compounds Containing Polyatomic Ions
    1. Polyatomic ions
    2. Formulas
        a. Electrical neutrality
        b. Crisscross
    3. Names (no suffix "ide")

IV. Covalent Bonds: Sharing Electrons

A. Covalent Bonding
    1. Sharing of electron pair(s) between nonmetal atoms
    2. Electrons shared to satisfy octet rule
    3. Crisscross rule

B. Multiple Covalent Bonds
    1. Triple bonds
    2. Double bonds

C. Naming Covalent Compounds
    1. Prefix + first nonmetal name - prefix + second nonmetal name stem + "ide"
    2. Number prefixes

V. Polar and Nonpolar Molecules: The Tug-of-War for Electrons

A. Partial Ionic and Partial Covalent Character

B. Nonpolar and Polar Molecules
    1. Electronegativity
    2. Nonpolar covalent bonds
    3. Polar covalent bonds
        a. Partial charges
        b. Dipoles

C. Net Molecular Polarity
    1. Dipoles that cancel one another

2. Dipoles that don't

VI. Forces Between Molecules: Why Everything Is Not a Gas
- A. Intermolecular Forces
- B. London Forces
  1. Temporary dipoles
  2. Exist between all chemical species
- C. Dipole-Dipole Interactions
- D. Hydrogen Bonds

VII. Solutions and Forces Between Molecules: Like Dissolves Like
- A. Definitions
  1. Solute, solvent, and solution
  2. Concentration
  3. Miscible and immiscible
- B. The Solution Process
  1. Energies involved
     - a. Overcoming attractions between solute molecules
     - b. Separating solvent molecules
     - c. Attractions formed between mixed molecules
  2. Like dissolves like
- C. Why Oil and Water Don't Mix
- D. Other Applications of Solubility
  1. Dry cleaning
  2. Fluoride toothpaste

Objectives

After you read and study the chapter [and the sections in brackets], you should be able to:

[R] 1. Write the charge on the ion (if any) formed from any atom according to the octet rule. [Section 4.1; Questions 6, 7, 14, 15, 16]

[C] 2. Predict whether any given compound is predominantly ionic or covalent. [Section 4.1; Questions 1, 2, 4, 5]

[M] 3. Associate the names, formulas, and charges of each polyatomic ion in Figure 4.11. [Section 4.3]

[R] 4. Write the formula of the compound formed from any given metal and nonmetal or polyatomic ion or from any two given nonmetals according to the octet rule. [Sections 4.3 and 4.4; Questions 5, 8, 9, 10]

[R] 5. Name any compound between two elements or polyatomic ions given its formula. [Sections 4.3 and 4.4; Question 5]

[R] 6. Predict whether a given molecule is polar or nonpolar, given its shape and formula. [Section 4.5; Questions 12, 17]

[R] 7. Determine, from the formula of a given substance, whether it has (a) dipole-dipole interactions, (b) hydrogen bonds, or (c) London forces between its molecules. [Section 4.6; Questions 12, 13]

[R] 8. Determine whether or not two given substances will mix to form a solution. [Section 4.7; Questions 18, 19, 20, 21]

Practice

Objective 1: Write the charge on the ion (if any) formed from any atom according to the octet rule.

Rule: Atoms gain or lose electrons to achieve the same number of valence electrons as the nearest noble gas.

Procedure: Locate the element on the periodic chart, determine its group number, and consult Figure 4.4.

Examples:

1. The charges on the ions formed from Sb, Ga, Cl, Xe, and Ca are 3-, 3+, 1-, 0 (i.e., no ion forms), and 2+, respectively.

Drill:

1. Give the charge, if any, on the ions formed from a) Se, b) Li, c) Br, d) Po, e) Ne, and f) H
2. An unknown element is found to form 1- ions. Is it likely to be an alkali metal, an alkaline earth, a transition element, a halogen, or a noble gas?

Objective 2: Predict whether any given compound is predominantly ionic or covalent.

Ionic compounds contain metals and/or polyatomic ions. **Covalent compounds** consist entirely of nonmetals and contain neither metals nor polyatomic ions.

Examples:

1. PbS_2 is IONIC because lead is a metal.
2. CS_2 is COVALENT because both carbon and sulfur are nonmetals.
3. $NaNO_3$ is IONIC both because Na is a metal and because nitrate is a polyatomic ion.
4. NH_4Cl is IONIC despite the fact that it consists only of nonmetals because it contains the polyatomic NH_4^{1+} ion. Look out for compounds containing ammonium ions; they are the only tricky case.

Drill:

1. Tell whether the following are predominantly Ionic or Covalent:

a)	N_2O_4	I	C
b)	$C_6H_{12}O_6$	I	C
c)	PbO_2	I	C
d)	SF_6	I	C
e)	$NaC_2H_3O_2$	I	C
f)	$(NH_4)_2S$	I	C
g)	$FeCl_3$	I	C
h)	$AlNaCO_3(OH)_2$	I	C

Objective 3: Associate the names, formulas, and charges of each polyatomic ion in Figure 4.11.

Drill:

Cut out the flash cards, and memorize the material on them.

Objective 4: Write the formula of the compound formed from any given metal and nonmetal or polyatomic ion or from any two given nonmetals according to the octet rule.

Rule: Atoms combine in a compound either by sharing or by transferring electrons such that each atom in the compound obeys the octet rule.

Procedure: To find the formula of a compound between two given elements or polyatomic ions:

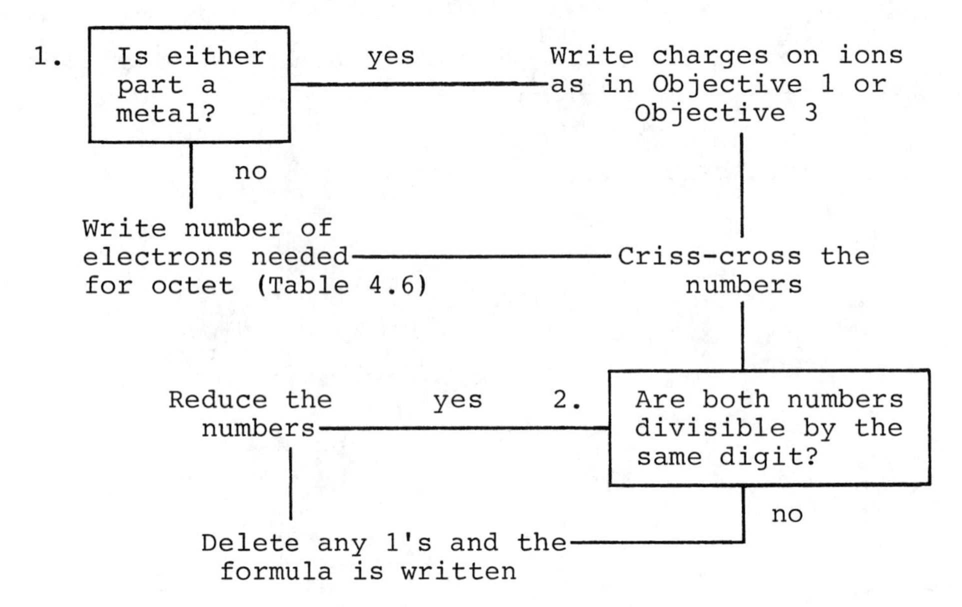

Examples:

1. The formula of a compound between carbon and oxygen:
   Question 1: no => C^4O^2 => C_2O_4;
   Question 2: yes => C_1O_2 => CO_2
   Note that this generates the formula of just one carbon-oxygen compound, carbon dioxide. Other compounds may or may not exist. In this case, carbon monoxide does, and its formula, CO, cannot be predicted by the rule.
2. The formula of a compound between calcium and chlorine:
   Question 1: yes => $Ca^{2+}Cl^{1-}$ => Ca_1Cl_2
   Question 2: no => $CaCl_2$
3. The formula of aluminum nitrate:
   Question 1: yes => Al^{3+} $NO_3{}^{1-}$ => $Al_1(NO_3)_3$
   Question 2: no => $Al(NO_3)_3$
   Note that you needed to know the formula and charge on the nitrate ion (Objective 3) and that you needed to place it in parentheses to get the correct formula. ($AlNO_{33}$ implies the silly notion that thirty-three

Flash Cards for Objective 3

ammonium ion	hydroxide ion
nitrate ion	perchlorate ion
cyanide ion	carbonate ion
sulfate ion	sulfite ion
phosphate ion	

OH^{1-}

NH_4^{1+}

ClO_4^{1-}

NO_3^{1-}

CO_3^{2-}

CN^{1-}

SO_3^{2-}

SO_4^{2-}

PO_4^{3-}

oxygens are combined with one nitrogen.)

4. The formula of barium sulfate:

   Question 1: yes => Ba^{2+} SO_4^{2-} => $Ba_2(SO_4)_2$

   Question 2: yes => $Ba_1(SO_4)_1$ => $BaSO_4$

   Here there is only one sulfate; therefore, it does not need parentheses.

Drill:

1. Write the formula of a compound between:
   a) sodium and oxygen
   b) chlorine and bromine
   c) lead and iodine
   d) magnesium and nitrogen
   e) arsenic and sulfur
   f) phosphorus and fluorine
2. What is the formula for each of the following?
   a) potassium phosphate
   b) strontium cyanide
   c) tin carbonate
   d) calcium hydroxide
   e) ammonium chloride
3. Which one of the following formulas is NOT correct?
   a) K_2Se b) GaSe c) CSe_2 d) Al_2Se_3

Objective 5: Name any compound between two elements or polyatomic ions given its formula.

Rule: If it is a metal-nonmetal compound, the name is:

name of metal + (charge) name-stem of nonmetal + "ide"

If it is a metal-polyatomic-ion compound, the name is:

name of metal + (charge) name of ion

If it is a nonmetal-nonmetal compound, the name is:

number prefix + name of 1st nonmetal number prefix + name-stem of 2nd nonmetal + "ide"

The name-stems are listed in Table 4.4; the number prefixes are found in Table 4.7; the polyatomic ions are in Figure 4.11.

Procedure: To name a compound from its formula:

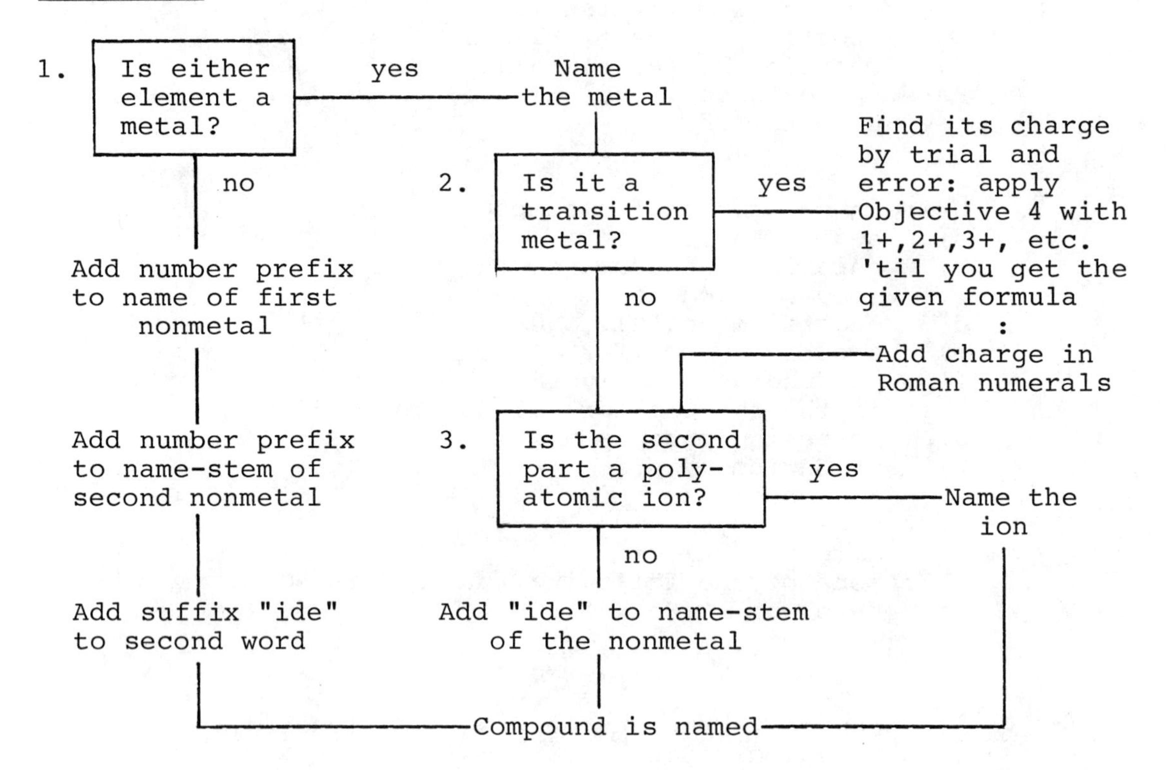

Examples:

1. The name of K_2S:
   Question 1: yes => potassium
   Question 2: no
   Question 3: no => sulf-ide => potassium sulfide
2. The name of NO_2:
   Question 1: no => mono-nitrogen (The prefix, mono, is optional; it is always correct either to include it or leave it out.) => di-ox => diox-ide => nitrogen dioxide or mononitrogen dioxide
3. The name of FeCl3:
   Question 1: yes => iron
   Question 2: yes => if iron is 1+, the formula would be Fe_1Cl_1; 2+: Fe_1Cl_2; 3+: Fe_1Cl_3; thus, the charge must be 3+ => iron(III)
   Question 3: no => chlor-ide => iron(III) chloride
4. The name of $Al(NO_3)_3$:
   Question 1: yes => aluminum
   Question 2: no
   Question 3: yes => nitrate => aluminum nitrate

5. The name of B_5H_9:
   Question 1: no => penta-boron => nona-hydr => nonahydr-ide => pentaboron nonahydride
6. The name of $CuSO_4$:
   Question 1: yes => copper
   Question 2: yes => 1+: $Cu_2(SO_4)_1$; 2+: $Cu_1(SO_4)_1$ => copper(II)
   Question 3: yes => sulfate => copper(II) sulfate

Drill:

1. Name the following:

a)	H_2O	f)	$Ca_3(PO_4)_2$
b)	$MnBr_2$	g)	HgO
c)	N_2O_4	h)	B_2H_6
d)	$Mg(CN)_2$	i)	PBr_3
e)	TiS_2	j)	IF_7

Objective 6: Predict whether a given molecule is polar or nonpolar, given its shape and formula.

Rule: A molecule is polar unless it is made up of a single element or it is so symmetrical that its bond polarities cancel out (i.e., perfectly straight, triangular, or tetrahedral).

Procedure: To predict whether a compound is polar or not:

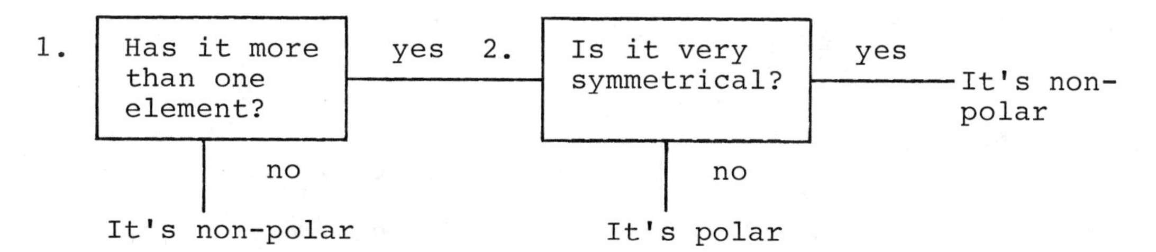

Examples:

1. S_8 (puckered ring shape):
   Question 1: no => non-polar
2. NO_2 (bent or angular shape):
   Question 1: yes; question 2: no => polar
3. SO_2 (straight or linear shape):
   Question 1: yes; question 2: yes => non-polar

Drill:

1. Distinguish the Polar from the Non-polar molecules.

a)	BH_3 (flat trianglar)	**P**	**N**
b)	N_2 (straight)	**P**	**N**
c)	NH_3 (umbrella shape)	**P**	**N**
d)	CH_4 (tetrahedral)	**P**	**N**

e)	H_2S (bent)	P	N
f)	P_4 (pyramidal)	P	N
g)	ClF_3 (T-shaped)	P	N
h)	HBr (straight)	P	N

Objective 7: Determine, from the formula of a given substance, whether it has (a) dipole-dipole interactions, (b) hydrogen bonds, or (c) London forces between its molecules.

Rule: Non-polar molecules have London forces between them. Polar molecules have dipole-dipole interactions between them unless the molecules contain hydrogen bound to fluorine, oxygen, or nitrogen. Then they have hydrogen bonds (mnemonic: H-FON bonds) between them.

Procedure: To predict the forces between molecules of a given substance:

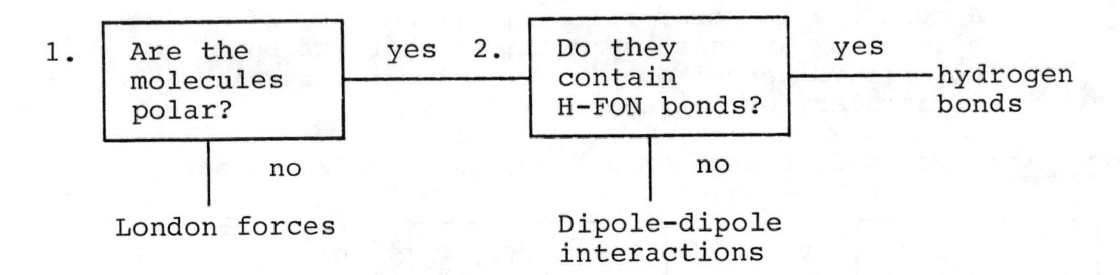

Examples:

1. H_2O (bent or angular shape):
   Question 1: yes; question 2: yes=> hydrogen bonds
2. TeH_2 (bent or angular shape):
   Question 1: yes; question 2: no => dipole-dipole
3. BeH_2 (straight or linear shape):
   Question 1: no => London forces

Drill:

1. Tell whether the following molecules have Dipole-dipole interactions, Hydrogen bonds, or London forces between them:

a)	BH_3 (flat trianglar)	D	H	L
b)	N_2 (straight)	D	H	L
c)	NH_3 (umbrella shape)	D	H	L
d)	CH_4 (tetrahedral)	D	H	L
e)	H_2S (bent)	D	H	L
f)	P_4 (pyramidal)	D	H	L
g)	ClF_3 (T-shaped)	D	H	L
h)	HBr (straight)	D	H	L

Objective 8: Determine whether or not two given substances will mix to form a solution.

Rule: Like dissolves like: if the molecules of two substances are alike in polarity (both polar or both non-polar) then they will mix to form a solution. Otherwise, they don't. If both substances are gases, however, it doesn't matter: they always mix regardless of polarity.

Procedure: To determine if two given substances will mix:

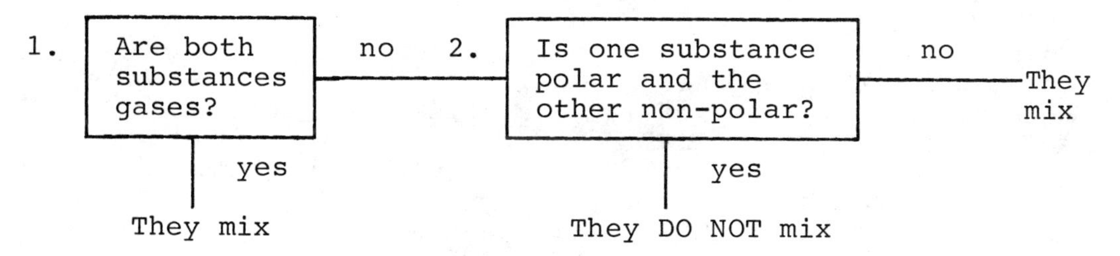

Examples:

1. Oil (non-polar) and water (polar):
   Question 1: no; question 2: yes => won't mix
2. Carbon dioxide gas (non-polar) and water vapor (polar):
   Question 1: yes => they mix
3. Butane (non-polar) and gasoline (non-polar):
   Question 1: no; question 2: no => they mix

Drill:

1. Which of the following pairs of substances are Miscible (mix together), and which are Immiscible (won't mix)?

a.	water (polar) and ammonia (polar)	M	I
b.	oil (non-polar) and salt (polar)	M	I
c.	carbon monoxide (polar) and air (non-polar)	M	I
d.	cleaning fluid (non-polar) and axle grease (non-polar)	M	I
e.	cleaning fluid (non-polar) and sticky stains (polar)	M	I
f.	diver's helium (non-polar) and compressed air (non-polar)	M	I
g.	peanut butter (non-polar) and jelly (polar)	M	I
h.	liquid nitrogen (non-polar) and liquefied petroleum gas (non-polar)	M	I

Self-Test

1. What would be the charge on a barium ion if it forms according to the octet rule?
2. Is RbF predominantly ionic or covalent?
3. SO_3 is a bent molecule. Is it polar or nonpolar?
4. HCl molecules are attracted to each other with hydrogen bonds. True or false?
5. CCl_4 is a nonpolar solvent. Will H_2 gas dissolve in it?
6. What is the formula of the compound between H and P?
7. What is the formula of copper(I) oxide?
8. Is I_2 polar or nonpolar?
9. When table salt is mixed with gasoline, the two form a solution. True or false?
10. Selenium, Se, takes on a _____ charge when it forms an ion according to the octet rule.
11. Write the formula of the phosphate ion.
12. Give the chemical name for $(NH_4)_2SO_4$.
13. The forces between molecules of CH_4 are
    a) London b) hydrogen c) dipole-dipole d) ionic
14. What is CN^-?
15. Name Mg_3N_2.
16. Write the formula of diboron hexahydride.

Answers

Objective 1:

1. a) 2-; b) 1+; c) 1-; d) 2-; e) 0 or none; f) 1+ or 1-
2. a halogen

Objective 2:

1. a) C; b) C; c) I; d) C; e) I; f) I; g) I; h) I

Objective 4:

1. a) Na_2O; b) ClBr; c) PbI_2; d) Mg_3N_2; e) As_2S_3; f) PF_3
2. a) K_3PO_4; b) $Sr(CN)_2$; c) $Sn(CO_3)_2$; d) $Ca(OH)_2$; d) NH_4Cl
3. b) should be Ga_2Se_3

Objective 5:

1. a) dihydrogen (mon)oxide; b) manganese(II) bromide; c) dinitrogen tetroxide; d) magnesium cyanide; e) titanium(IV) oxide; f) calcium phosphate; g) mercury(II) oxide; h) diboron hexahydride; i) (mono)phosphorus tribromide; j) (mono)iodine heptafluoride

Objective 6:

1. a) N; b) N; c) P; d) N; e) P; f) N; g) P; h) P

Objective 7:

1. a) L; b) L; c) H; d) L; e) D; f) L; g) D; h) D

Objective 8:

1. a) M; b) I; c) M; d) M; e) I; f) M; g) I; h) M

Self-Test:

1) 2+; 2) ionic; 3) polar; 4) false; 5) yes; 6) H_3P or PH_3; 7) Cu_2O; 8) nonpolar; 9) false; 10) 2-; 11) PO_4^{3-}; 12) ammonium sulfate; 13) a; 14) cyanide; 15) magnesium nitride; 16) B_2H_6

Evaluation:

If you missed more than one question in any of the following groups, you need to review the section indicated:

Question Groups	Section
1, 10	4.1
11, 14	4.3
2, 6, 7, 12, 15	4.3 and 4.4
3, 8	4.5
4, 13	4.6
5, 9	4.7

Chapter 5

Physical and Chemical Changes: Rearranging Matter

Outline

I. Physical States: Solids, Liquids, and Gases
   A. States of Matter
      1. Solid--definite volume and shape
      2. Liquid--definite volume but no definite shape
      3. Gas--no definite volume or shape
   B. Some Physical Properties of Gases
      1. Pressure and temperature: directly proportional
      2. Pressure and volume: inversely proportional
      3. Volume and temperature: directly proportional
   C. The Kinetic Molecular Theory
      1. Molecules in continuous, rapid, random motion which increases with temperature
      2. Motion (disruptive forces) opposes intermolecular forces (attractive forces)
      3. Physical states and intermolecular forces
         a. Solids: attractive forces dominate disruptive
         b. Liquids: attractive and disruptive forces balanced
         c. Gases: disruptive forces dominate attractive
      4. Kinetic molecular theory and characteristic properties of gases, liquids, and solids

II. Physical Changes: Melting and Boiling
   A. Melting Point and Freezing Point
      1. Melting point: temperature of switchover from solid to liquid
      2. Freezing point: same temperature; reverse switchover
      3. Characteristic properties of pure substances
   B. Vaporization and Vapor Pressure
      1. Vaporization: conversion from liquid to gas
      2. Condensation: reverse conversion
      3. Dynamic equilibrium: balance of two opposing processes
      4. Vapor pressure: pressure of gas at equilibrium over a liquid in a closed container
   C. Boiling Point
      1. Boiling point: temperature at which vapor pressure equals atmospheric pressure
      2. Normal boiling point: temperature at which vapor pressure equals sea-level pressure
   D. Boiling Point and Intermolecular Forces

III. Chemical Changes: What Is a Chemical Reaction?
   A. Chemical Reactions: Making and Breaking Bonds

1. Reactants and products
2. Evidences of chemical change
    a. Gas forms
    b. Odors or colors change
    c. Precipitate forms
    d. Flames
    e. Relatively large amounts of heat or light involved
3. Chemical vs. physical
    a. Chemical: structural rearrangement of atoms, ions, or molecules
    b. Physical: all atoms, ions, and molecules retain their identities

B. Chemical Equations

C. Balancing Chemical Equations
1. Upholding law of conservation of mass
2. Coefficients vs. subscripts

D. Limitations of Balanced Equations
1. Reaction may not actually happen
2. Does not indicate how reaction occurs

IV. Ten Important Chemical Reactions: Some Useful Chemical Sentences

A. Reactions 1 and 2: Burning Fossil Fuels

B. Reactions 3 and 4: Nitrogen Oxides, Chemical Villains in Our Air

C. Reactions 5, 6, and 7: Sulfur Dioxide, Valuable Resource or Another Villain?

D. Reaction 8: The Manufacture of Ammonia

E. Reaction 9: Photosynthesis in Plants

F. Reaction 10: Respiration in Cells

V. The Mole Concept: Some Simple Arithmetic of Chemical Reactions

A. Avogadro's Number: Counting Structural Particles of Elements and Compounds
1. Standard bunch or number
2. 6.02×10^{23}

B. Avogadro's Number and the Mole Concept

C. The Mole Concept and the Molar Mass of Elements and Compounds
1. Molar mass of an element
2. Molar mass of a compound

D. Using the Mole and Molar Mass Concepts to Make Calculations from Balanced Chemical Equations

E. Mole-to-Mole Calculations

F. Mole-to-Mass Calculations

G. Mass-to-Mass Calculations

Objectives

After you read and study the chapter [and the sections in brackets], you should be able to:

[C] 1. Use the kinetic molecular theory to interpret:
- a. the differences among solids, liquids, and gases [Section 5.1; Question 1]
- b. the relations among pressure, volume and temperature of gases [Section 5.1; Questions 2, 3, 5]
- c. the phenomena of boiling point, melting point, and vapor pressure. [Section 5.2; Questions 4,5]

[C] 2. Distinguish between chemical and physical changes, given a description of the phenomena. [Section 5.3; Questions 6, 10]

[C,R] 3. Classify chemical equations as balanced or not balanced. [Section 5.3; Questions 7, 8, 9, 11]

[M] 4. Recall the ten important chemical reactions in Section 5.4, and write their balanced equations. [Section 5.4; Question 12]

[R] 5. Interconvert reaction quantities (masses and/or moles) using chemical arithmetic [Section 5.5; Questions 13, 14, 15, 16, 17, 18]

Practice

Objective 1a: Use the kinetic molecular theory to interpret the differences among solids, liquids, and gases.

The molecules in **solids** have strong attractive forces (electrostatic, London, dipole-dipole, or hydrogen) but weak disruptive forces (heat); in **gases** the situation is reversed: strong disruptive forces and weak attractive forces; in **liquids** the forces are balanced. See Objective 3b of Chapter 1.

Examples:

1. Would you expect RbF to a gas, liquid, or solid at room temperature? SOLID; as with all metal-nonmetal (ionic) compounds, its attractive forces are electrostatic, the strongest of all. Thus, it's reasonable to assume that they dominate at room temperature.
2. Would NF_3 be a gas, liquid, or solid at a thousand degrees? GAS; nonmetal-nonmetals have weak London forces as their only intermolecular attractions, and at 1000° the disruptive forces are large.

Drill:

1. Tell whether the following out-of-the-ordinary molecules are Gases, Liquids, or Solids at the given temperature:
   a) B_2H_6 at room temperature G L S
   b) $BaCl_2$ at 0° G L S
   c) C_8H_{18} at room temperature G L S

Objective 1b: Use the kinetic molecular theory to interpret the relations among pressure, volume and temperature of gases.

As the temperature increases in a gas, the molecules move faster, and they collide more violently with the walls of the container. Thus, both volume and pressure tend to increase with temperature.

In smaller volumes, the gas molecules collide more often with the walls, and the pressure increases. At larger volumes, the opposite holds. Pressure and volume have an opposite or see-saw relation.

Examples:

1. What happens to the volume if the pressure is increased? It decreases. Squeezing on a volume of gases makes them contract into a smaller volume.
2. What happens to the temperature if the volume decreases but the pressure stays the same. At the same pressure, the molecules continue to collide with the walls at the same force. But at smaller volume, they do so more often. Thus each collision must be weaker than before or the pressure would change. All this means the temperature must drop.

Drill:

1. Tell what happens to the specified property under the given conditions. Does it Increase, Decrease, or Stay the same?
   a) Pressure at greater temperature and same volume I D S
   b) Volume at lesser temperature and same pressure I D S
   c) Volume at twice the temperature and twice the pressure I D S
   d) Temperature at increased volume and same pressure I D S
   e) Pressure at same temperature and greater volume I D S

Objective 1c: Use the kinetic molecular theory to interpret the phenomena of boiling point, melting point, and vapor pressure.

As the temperature increases in a solid, there comes a point at which the attractive and disruptive forces balance. This is the **melting point.** The solid then changes to liquid.

As the temperature continues to rise in the liquid, there comes a point at which the disruptive forces overcome the attractive forces. This is the **boiling point,** and the liquid changes to gas.

At any temperature and in either solids or liquids, some individual molecules gain enough energy to escape into the gas phase. Their collective collisions against the walls of the container constitute a liquid's (or solid's) **vapor pressure.**

Examples:

1. a) Why does an open bottle of perfume evaporate? Because the molecules, one by one, gain enough energy to escape. b) Why won't the perfume evaporate from a well-sealed bottle? Because of the confined space inside the bottle. As more molecules evaporate, other previously-evaporated molecules re-enter the liquid; thus, after a while, these two opposing processes cancel, and the amount of perfume in the gas phase ceases to change. Its vapor pressure is established.
2. Why is oxygen a gas? Because of the weak forces of attraction between its molecules. The disruptive forces are comparatively so great at room temperature, that the temperature is above both oxygen's melting point and boiling point. Oxygen at room temperature has both melted and boiled away into gas.

Drill:

1. Why is the boiling point of table salt so high? ($1413^{o}C$ compared to water's $100^{o}C$)
2. Is the vapor pressure of hospital ether higher or lower than that of water at room temperature?

Objective 2: Distinguish between chemical and physical changes, given a description of the phenomena.

Chemical changes involve the production of a new substance. Clues that indicate this production include bubbles of gas, new odors or colors, solid precipitates, flames, and relatively large amounts of heat or light.

Physical changes retain the identity of the substance. Remember that these changes include all changes of state and most dissolvings.

Examples:

1. Mixing hydrogen and oxygen in a balloon. PHYSICAL CHANGE. Both gases retain their identities in the balloon.
2. Igniting a hydrogen-oxygen mixture. CHEMICAL CHANGE. Large production of heat and light; new substance, water, appears.
3. Bubbles of gas appear when acid is added to a baking soda solution. CHEMICAL CHANGE. The gas bubbles are a new substance, carbon dioxide.
4. Bubbles of gas appear when water is boiled. PHYSICAL CHANGE. The gas bubbles are still water in this change of state.

Drill:

1. Classify the following as Chemical or Physical changes:

a)	garbage gets smelly	C	P
b)	dust settles	C	P
c)	autumn leaves turn to gold	C	P
d)	food is digested in the stomach	C	P
e)	beer foams when it is poured	C	P
f)	snow melts	C	P
g)	soles of shoes become thin	C	P
h)	peeled bananas turn brown	C	P
i)	sawdust burns	C	P

Objective 3: Classify chemical equations as balanced or not balanced.

A **balanced chemical equation** results when the number of atoms of each element in the reactants (on the left of the arrow) equals the number of atoms of the same element in the products (on the right of the arrow).

Rule: To count the number of atoms of each element on one side of the equation, (1) consider each species that contains the element. Then, (2) multiply the coefficient (the big numeral in front of the species) by the subscript of the element (the small numeral to the right and below). If the element is enclosed by parentheses, (3) multiply the previous product by the subscript outside the parentheses. Finally, (4) add up the contributions from each species to get the final tally.

Procedure:

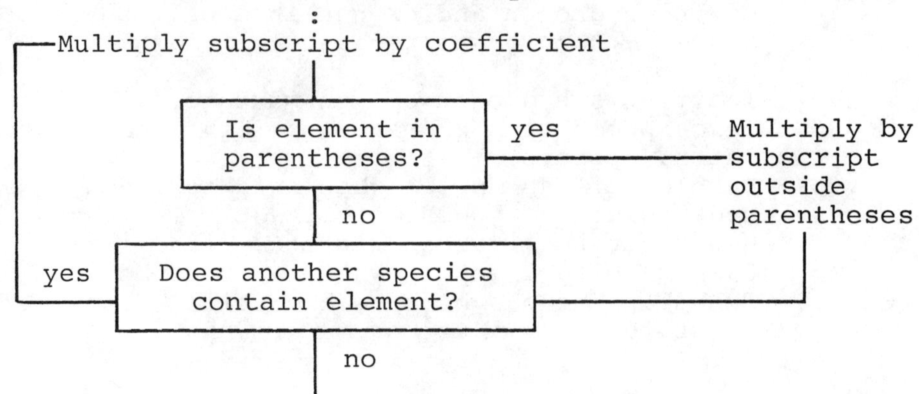

Examples:

1. $2\ H_2O_2 \rightarrow H_2O + O_2$
   Tally reactants: **H:** 2 x 2; no; no; 4 H's as reactants. **O:** 2 x 2; no; no; 4 O's
   Tally products: **H:** 2 x 1 (when no coefficient or subscript is written, assume it is 1); no; no; 2 H's as products. **O:** 1 x 1 (in the H_2O); no; yes; 2 x 1 (in the O_2); no; no; 1 + 2 = 3 O's as products.
   Neither the H's nor the O's balance; therefore, the equation is NOT BALANCED.
2. $ZnCl_2 + 2\ NaOH \rightarrow Zn(OH)_2 + 2\ NaCl$
   Tally reactants: **Zn:** 1 x 1; no; no; 1 Zn. **Cl:** 2 x 1; no; no; 2 Cl's. **Na:** 1 x 2; no; no; 2 Na's. **O:** 1 x 2; no; no; 2 O's. **H:** 1 x 2; no; no; 2 H's
   Tally products: **Zn:** 1 x 1; no; no; 1 Zn. **O:** 1 x 1; yes; 2 x 1; no; 2 O's. **H:** 1 x 1; yes; 2 x 1; no; 2 H's. **Na:** 1 x 2; no; no; 2 Na's. **Cl:** 1 x 2; no; no; 2 Cl's.
   The Zn's, Cl's, Na's, O's, and H's all balance without exception; therefore the equation is BALANCED.
3. $Fe_2O_3 + 2\ CO \rightarrow 2\ Fe + 2\ CO_2$
   Tally reactants: **Fe:** 2 x 1; no; no; 2 Fe's. **O:** 3 x 1; no; yes; 1 x 2; no; no; 5 O's. **C:** 1 x 2; no; no; 2 C's.
   Tally products: **Fe:** 1 x 2; no; no; 2 Fe's. **C:** 1 x 2; no; no; 2 C's. **O:** 2 x 2; no; no; 4 O's.
   Although both the Fe's and the C's balance perfectly and the O's are only off by one, the whole equation is NOT BALANCED.

Drill:

1. Tell whether each of the following equations is Balanced or Not Balanced:

a)	NH_4NO_3 --> N_2O + 2 H_2O	B	N
b)	$CuCO_3$ + 2 HCl --> $CuCl_2$ + CO_2 + H_2O	B	N
c)	C_4H_{10} + 7 O_2 --> 4 CO_2 + 5 H_2O	B	N
d)	3 PF_2Br --> 2 PF_3 + PBr_3	B	N
e)	I_2 + H_2S --> S + 2 HI	B	N
f)	Br_2 + H_2O --> 2 HBr + HBrO	B	N
g)	$FeCl_3$ + 3 NaOH --> $Fe(OH)_3$ + 3 NaCl	B	N
h)	$Cr_2(SO_4)_3$ + 3 KOH --> $Cr(OH)_3$ + 3 K_2SO_4	B	N

Objective 4: Recall the ten important chemical reactions in Section 5.4, and write their balanced equations.

Drill:

Cut out the flash cards, and memorize the material on them.

Objective 5: Interconvert reaction quantities (masses and/or moles) using chemical arithmetic.

Rule: (1) Determine the units of the answer. (When the problem is solved, what units will the final quantity have?) (2) Start with the given quantity. (Which number among all those given in the problem is not the ratio of, or the relation between, two quantities?) (3) Plot a course to the solution by going from the Step 2 units to the Step 1 units in the boxes below. (How many conversion factors will you need altogether?)

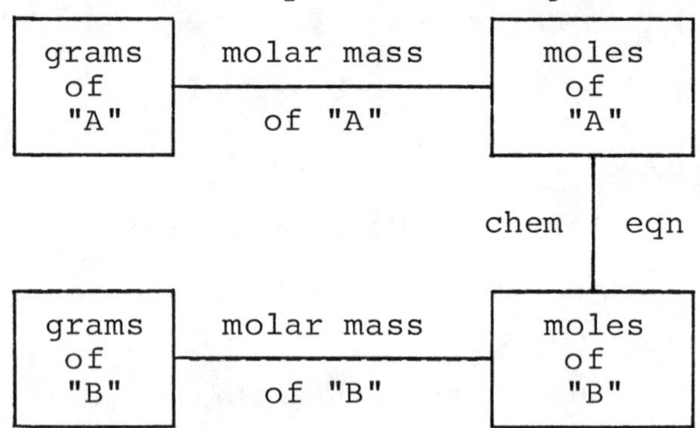

(4) Multiply by each conversion factor--positioning it either right-side-up or upside down--so that the previous unit cancels out. (Which way will put the previous unit in the denominator?) (5) Continue with Step 4 until you arrive at the units of the answer. (When will you know to stop this process?)

Procedure:

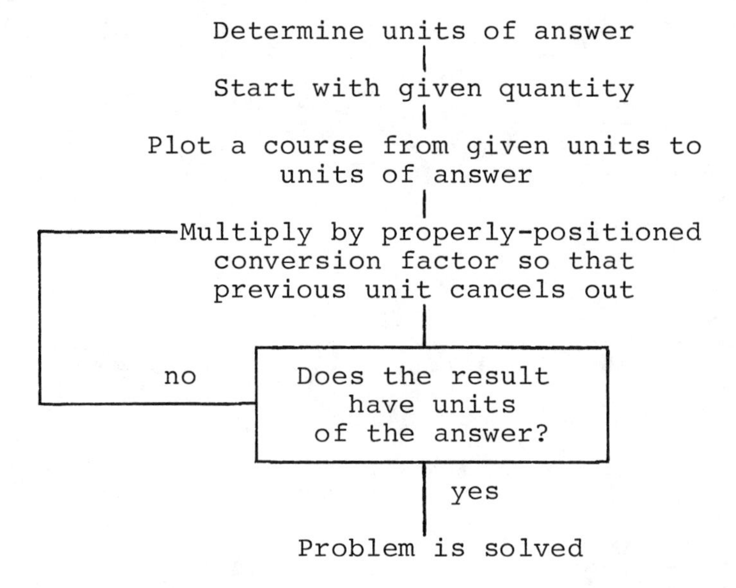

Flash Cards for Objective 4

Reaction 1 The Burning of Carbon	Reaction 2 The Burning of Octane
Reaction 3 The Formation of Nitric Oxide	Reaction 4 The Formation of Nitrogen Dioxide
Reaction 5 The Formation of Sulfur Dioxide	Reaction 6 The Formation of Sulfur Trioxide
Reaction 7 The Synthesis of Sulfuric Acid	Reaction 8 The Synthesis of Ammonia
Reaction 9 Photosynthesis	Reaction 10 Respiration

$$2\ C_8H_{18} + 25\ O_2 \rightarrow 16\ CO_2 + 18\ H_2O$$

$$C + O_2 \rightarrow CO_2$$

$$2\ NO + O_2 \rightarrow 2\ NO_2$$

$$N_2 + O_2 \rightarrow 2\ NO$$

$$2\ SO_2 + O_2 \rightarrow 2\ SO_2$$

$$S + O_2 \rightarrow SO_2$$

$$N_2 + 3\ H_2 \rightarrow 2\ NH_3$$

$$SO_3 + H_2O \rightarrow H_2SO_4$$

$$C_6H_{12}O_6 + 6\ O_2 \rightarrow 6\ CO_2 + 6\ H_2O$$

$$6\ CO_2 + 6\ H_2O \rightarrow C_6H_{12}O_6 + 6\ O_2$$

Examples:

1. How many moles of H_2O are there in 54.0 grams of H_2O?

   If we let H_2O be substance "A", then the units of the answer are moles of "A". Given are 54.0 g "A". The conversion factor we need is the molar mass of H_2O, or (2 x mass of H) + (mass of O) = (2 x 1.0) + (16.0) = 18 g H_2O / 1 mol H_2O.

$$? \text{ mol } H_2O = 54.0 \text{ g } H_2O \times \frac{1 \text{ mol } H_2O}{18.0 \text{ g } H_2O} = 3.0 \text{ mol } H_2O$$

   The conversion factor needed to be upside down. When grams H_2O cancel, we are left with the units of the answer. And so we are done. (The results would have been identical had we chosen "B" as H_2O.)

2. How many grams of O_2 can be produced by decomposing 1.0 mol of $KClO_3$ according to the reaction:

$$2\ KClO_3 \text{ --> } 2\ KCl + 3\ O_2$$

   There are two substances involved. Let's call O_2 substance "A" and $KClO_3$ substance "B". (It works equally well to name them the opposite way; try it out.) We are looking for grams of "A", and we have 1.0 mol of "B". According to the boxes, we'll need to do two conversions: mol "B" to mol "A", then mol "A" to g "A". The conversion factors are, first, from the chemical equation, 2 mol $KClO_3$ / 3 mol O_2, and, second, the molar mass of O_2 or (2 x 16.0) = 32.0 g O_2 / 1 mol O_2.

$$? \text{ g } O_2 = 1.0 \text{ mol } KClO_3 \times \frac{3 \text{ mol } O_2}{2 \text{ mol } KClO_3} \times \frac{32.0 \text{ g } O_2}{1 \text{ mol } O_2}$$
$$= 48.0 \text{ g } O_2$$

   Both conversion factors were right-side up.

3. Silver tarnishes when it reacts with sulfur to give black silver sulfide, Ag_2S. How many grams of Ag are consumed by 4.0 grams S in the reaction:

$$2\ Ag + S \text{ --> } Ag_2S$$

   We must find g Ag starting with 4.0 g S. Since there is no line connecting g "A" with g "B" directly, we have to do a three-step process: g S to mol S, then mol S to mol Ag, then mol Ag to g Ag. The conversion factors are 32.0 g S / 1 mol S, 2 mol Ag / 1 mol S, and 108 g Ag / 1 mol Ag, respectively.

$$? \text{ g Ag} = 4.0 \text{ g S} \times \frac{1 \text{ mol S}}{32.0 \text{ g S}} \times \frac{2 \text{ mol Ag}}{1 \text{ mol S}} \times \frac{108 \text{ g Ag}}{1 \text{ mol Ag}}$$

$$= 27.0 \text{ g Ag}$$

Note that the first conversion factor needed to be turned upside down, but the other two remained upright. Note also that no conversion factor involving Ag_2S was necessary.

Drill:

1. Convert 8.0 mol SO_2 to grams of SO_2.
2. How many moles of $CaCO_3$ are there in 25.0 g?
3. What is the mass in grams of 1.25 mol $(NH_4)_2SO_4$?
4. A bottle contains 0.40 mol of wood alcohol, CH_3OH. How much does the alcohol weigh in grams?
5. How many moles of water are produced when 2.0 mol of NH_3 react according to the equation

   $4\ NH_3 + 5\ O_2 \text{ --> } 4\ NO + 6\ H_2O$

6. Calculate the number of moles of Fe_2O_3 are necessary to produce 7.0 mol CO_2 according to the equation:

   $3\ Fe_2O_3 + CO \text{ --> } 2\ Fe_3O_4 + CO_2$

7. How many moles of Ca_3P_2 will react with 3.0 mol of water in the reaction:

   $Ca_3P_2 + 6\ H_2O \text{ --> } 3\ Ca(OH)_2 + 2\ PH_3$

8. Determine the number of moles of sulfur that are produced along with 4.2 mol H_2O in the reaction:

   $2\ H_2S + SO_2 \text{ --> } 2\ H_2O + 3\ S$

9. How many grams of O_2 will react with 1.0 mol of NH_3 in the reaction of Question 5?
10. If 2.0 mol of CO react as in Question 6, how many grams of CO_2 result?
11. Convert 0.68 g PH_3 to the number of moles of Ca_3P_2 necessary to produce it by the reaction in Question 7.
12. How many grams of SO_2 will it take to create 5.0 mol H_2O in the reaction of Question 8?
13. In the reaction of Question 7, how many grams of $Ca(OH)_2$ are produced if 21.6 g of PH_3 are a by product?
14. If the reaction in Question 5 consumes 108 g of O_2, how many grams of NO are produced?
15. What mass of H_2S is necessary to react completely with 1280 g of SO_2 according to the reaction in Question 8?
16. Convert 15.0 g of Fe_2O_3 to grams of Fe_3O_4 using the equation in Question 6

Self-Test

1. Does an increase in temperature on a sample of gas make its volume tend to increase or decrease?
2. True or false, cracking open an egg is a chemical change.
3. Is $2\ Na + H_2O \rightarrow 2\ NaOH + H_2$ a balanced equation?
4. Write the equation for the synthesis of ammonia.
5. As ice forms from water, do the H_2O molecules speed up or slow down?
6. In $MgO + Si \rightarrow SiO_2 + Mg$, the element out of balance is
   a) Mg b) O c) Si d) none, they all balance
7. The temperature at which the disruptive forces of heat just balance the attractive forces in a solid is called what?
8. What forces predominate in gases, attractive or disruptive?
9. Write the equation for our respiration of oxygen in the cells.
10. A pressure gauge says that the pressure dropped when a constant amount of gas when the temperature is changed in a rigid container. Was the temperature raised or lowered?
11. When a balloon bursts and all the air leaks out, is that a chemical or physical change?
12. Write the equation for the burning of coal.
13. When water evaporates at room temperature, is that boiling?
14. For the reaction $2\ H_2O_2 \rightarrow 2\ H_2O + O_2$ calculate:
    a) The number of moles of H_2O_2 in 68.0 g.
    b) The grams of O_2 formed when 0.10 mol H_2O_2 decomposes.
    c) The moles of H_2O_2 necessary to produce 8 mol of O_2.
    d) The grams of water produced along with 4.0 g O_2.

Answers

Objective 1a:

1. a) G; b) S; c) L

Objective 1b:

1. a) I; b) D; c) S; d) I; e) D

Objective 1c:

1. Because the attractive forces in this ionic compound are so great.
2. Higher; it evaporates more easily because its intermolecular forces are smaller.

Objective 2:

1. a) C; b) P; c) C; d) C; e) P; f) P; g) P; h) C; i) C

Objective 3:

1. a) B; b) B; c) N; d) B; e) B; f) N; g) B; h) N

Objective 5:

1) 512 g; 2) 0.25 mol; 3) 165 g; 4) 12.8 g; 5) 3.0 mol; 6) 21 mol; 7) 0.5 mol; 8) 6.3 mol; 9) 40 g; 10) 88 g; 11) 0.01 mol; 12) 175 g; 13) 70.5 g; 14) 81.0 g; 15) 1360 g; 16) 14.5 g

Self-Test:

1) increase; 2) false; 3) no; 4) $3\ H_2 + N_2 \text{ --> } 2\ NH_3$; 5) slow down; 6) b; 7) melting point; 8) disruptive; 9) $C_6H_1{}^2O_6 + 6\ O_2 \text{ --> } 6\ CO_2 + 6\ H_2O$; 10) lowered; 11) physical; 12) $C + O_2 \text{ --> } CO_2$; 13) no; 14a) 2 mol H_2O_2; 14b) 0.16 g O_2; 14c) 16 mol H_2O_2; 14d) 4.5 g H_2O

Evaluation:

If you missed more than one question on the self-test in any of the following groups, you need to review the section indicated:

Question Groups	Section
1, 5, 10	5.1
7, 8	5.2
2, 3, 6, 11	5.3
4, 9, 12	5.4
14a, 14b, 14c, 14d	5.5

Chapter 6

Acid-Base and Oxidation-Reduction Reactions: Transferring Protons and Electrons

Outline

I. Acids, Bases, and Acid-Base Reactions: Pass the Protons
    A. Acids
        1. Donate hydrogen ions or protons to water
        2. Common properties
            a. Formulas start with H
            b. Sour or tart taste
            c. Turn litmus red
        3. Acid strength
            a. Strong: react completely with water to produce hydronium ions
            b. Weak: react only slightly
        4. Not all acids are corrosive
    B. Bases
        1. Produce hydroxide ions in water, or accept protons
        2. Base strength
            a. Strong: release all of its hydroxides to water
            b. Weak: release relatively few hydroxides
                (1) have low solubility
                (2) react only slightly
        3. Common properties
            a. Formulas end in OH
            b. Bitter taste
            c. Slippery, soapy feel
            d. Turn litmus blue
    C. Acid and Base Strengths: The pH Scale
        1. Solutions
            a. Acid: more hydronium ions than hydroxide
            b. Base: more hydroxide ions than hydronium
            c. Neutral: equal concentrations
        2. pH scale: measure of hydronium ion concentration
            a. below 7: acidic
            b. exactly 7: neutral
            c. above 7: alkaline
    D. Acid-Base Reactions
        1. Proton transfer reactions
        2. Examples
II. Some Useful Acid-Base Reactions: Reducing Excess Stomach Acidity, Improving Lakes and Soils, and Baking Bread
    A. Antacids: Reducing Excess Stomach Acidity
        1. Hyperacidity
        2. Common antacid ingredients
            a. $Mg(OH)_2$ and $Al(OH)_3$

b. $CaCO_3$
c. $NaHCO_3$
d. $AlMg(OH)_5$ and $AlNaCO_3(OH)_2$
3. Properties of ideal antacid
B. Reducing Excess Acidity in Lakes and Soils
1. Acid rain
2. Neutralizing acidic or basic soils
C. Baking Bread

III. Oxidation, Reduction, and Oxidation-Reduction Reactions: Pass the Electrons
A. Oxidation and Reduction: The Loss or Gain of Electrons
1. Oxidation
a. The gain of oxygen atoms
b. The loss of electrons
2. Reduction
a. The loss of oxygen atoms
b. The gain of electrons
B. Oxidation-Reduction Reactions
1. Electron transfer (redox)
2. Oxidizing and reducing agents
3. Recognizing oxidation and reduction
a. Oxidation
(1) loss of electrons
(2) increase in electrical charge
(3) gain of oxygen
(4) loss of hydrogen
b. Reduction
(1) gain of electrons
(2) decrease in electrical charge
(3) loss of oxygen
(4) gain of hydrogen

IV. Some Useful Oxidation-Reduction Reactions: Bleaching, Water Purification, Stain Removal, and Photography
A. Bleaching
B. Water Purification
C. Removing Stains
D. Photography
1. Exposing and developing
2. Making prints
3. Color photography

V. Energy From Oxidation-Reduction Reactions: Batteries
A. Setting Up a Battery
B. Commercial Batteries
1. Common batteries
2. Fuel cells

Objectives

After you read and study the chapter [and the sections in brackets], you should be able to:

[C] 1. Identify acids and bases given:
   a. the ions they form in solution, or their formulas, tastes in foods, or litmus tests. [Section 6.1; Questions 1, 2, 3]
   b. the equation of a proton-transfer reaction. [Section 6.1; Question 7]

[R] 2. Compare the strengths of acids and bases using the pH scale. [Section 6.1; Questions 5, 6, 8, 9, 10]

[C] 3. Identify acid-base reactions involved in reducing excess stomach acidity, improving lakes and soils, and baking bread. [Section 6.2; Questions 4, 11, 12]

[C] 4. Classify the reactants in a given chemical reaction as:
   a. being oxidized, reduced, or neither of the two. [Section 6.3; Questions 13, 15]
   b. oxidizing agents, reducing agents, or neither of the two. [Section 6.3; Questions 14, 16]

[C] 5. Identify oxidation-reduction reactions involved in bleaching and stain removal, water purification, photography, and batteries. [Sections 6.4, 6.5; Questions 17, 18]

Practice

Objective 1a: Identify acids and bases given the ions they form in solution, or their formulas, tastes in foods, or litmus tests.

Acids produce H^+ ions in solution, have formulas that start with H, taste sour or tart in foods, and turn litmus red.
Bases produce OH^- ions in solution, have formulas that end in OH, taste bitter in foods, and turn litmus blue.
The identity of the ions in solution is the surest test for acids and bases, followed closely by the litmus test. The other clues, though generally reliable, are not infallible.

Examples:

1. A blue solution that contains H^+. ACID; (color is irrelevant unless litmus is present).
2. A bitter pill. BASE.

3. H_3BO_3 and $Mg(OH)_2$. ACID and BASE, respectively; (neither the number of H's at the beginning of a formula nor OH's at the end are important).

Drill:

1. Tell whether each of the following is an Acid or a Base:

a)	Sour cream	A	B
b)	$AlNaCO_3(OH)_2$	A	B
c)	Solution containing NH_4^+ and OH^- ions	A	B
d)	Litmus paper turns from red to blue	A	B
e)	Unsweetened chocolate	A	B
f)	A mixture of H^+ and NO_3^- ions	A	B
g)	$HC2H_3O_2$	A	B

Objective 1b: Identify acids and bases the equation of a proton-transfer reaction.

Acids donate protons (H^+ ions); that is, in going from reactant to product, they end up with fewer H's in their formulas.

Bases accept protons (H^+ ions); that is, in going from reactant to product, they end up with more H's in their formulas.

If a reaction involves a proton transfer, both an acid and a base must be present; neither can be found alone.

Examples:

1. $HBr + NH_3 \text{-->} Br^- + NH_4^+$. HBr is an ACID. (Comparing the before, HBr, with the after, Br^-, notice that it has lost an H). NH_3 is a BASE. (Starting as NH_3, it ends up with a extra H as NH_4^+)
2. $NH_2^- + H_2O \text{-->} NH_3 + OH^-$. NH_2^- gains an H and is the BASE. H_2O loses an H, and is the ACID.
3. $H_2PO_4^- + H_2O \text{-->} HPO_4^{2-} + H_3O^+$. Here the water gains an H and acts as the BASE. $H_2PO_4^-$ lose one H and is the ACID.

Drill:

1. In the following equations, circle the acid and underline the base:

a) $CH_3NH_2 + HCl \text{-->} CH_3NH_3^+ + Cl^-$

b) $H_2SO_4 + 2\ NaOH \text{-->} Na_2SO_4 + 2\ H_2O$

c) $H_2O + HS^- \text{-->} OH^- + H_2S$

d) $HC_2H_3O_2 + CO_3^{2-} \text{-->} C_2H_3O_2^- + HCO_3^-$

e) $HF + NH_3 \rightarrow F^- + NH_4^+$

f) $C_6H_5NH_2 + HSO_4^- \rightarrow C_6H_5NH_3^+ + SO_4^{2-}$

Objective 2: Compare the strengths of acids and bases using the pH scale.

Rule: In comparisons between two solutions, the one with the lower pH is the more acidic. The one with the greater pH is the more alkaline, or basic.
For a single solution, its pH is implicitly compared to that of pure water (a pH of 7). Thus, if the solution's pH is less than 7, it is acidic. If it is greater than 7, the solution is alkaline. With pH equal to 7, the solution is neutral, neither acidic nor alkaline.

Procedure:

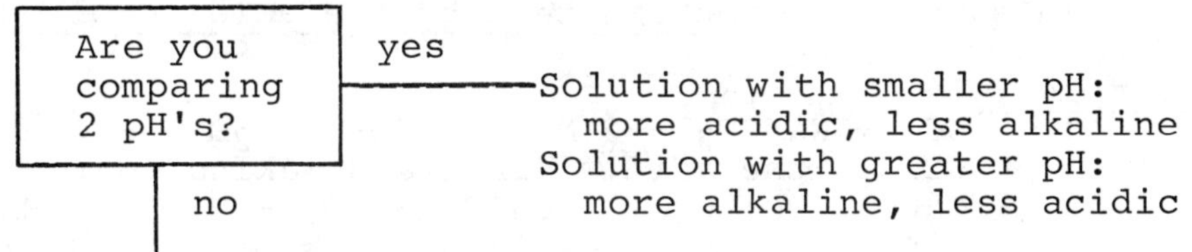

pH greater than 7: alkaline
pH equal to 7: neutral
pH less than 7: acidic

Examples:

1. An antacid solution has a pH of 9.2. No; pH>7; ALKALINE.
2. Which is more acidic, the antacid with a pH of 9.2 or a drain cleaner with a pH of 13.4? Yes; 9.2 < 13.4; the antacid is MORE ACIDIC relative to the drain cleaner even though it is still alkaline compared to water.

Drill:

1. Identify solutions having the following pH's as Acidic, Basic, or Neutral:

a)	6.99	A	B	N
b)	5.1	A	B	N
c)	11	A	B	N
d)	7.0	A	B	N
e)	1	A	B	N
f)	7.3	A	B	N

2. Which pH would indicate the more acidic solution in each of the following pairs?

a) 5.4 or 1.1
b) 13 or 10

c) 7.0 or 9.0
d) 6.35 or 8.71
e) 1 or 2
f) 2 or 14

Objective 3: Identify acid-base reactions involved in reducing excess stomach acidity, improving lakes and soils, and baking bread.

In **reducing excess stomach acidity,** the acid is always HCl reacting with an antacid base such as $Mg(OH)_2$, $Al(OH)_3$, $CaCO_3$, $NaHCO_3$, $AlMg(OH)_5$, or $AlNaCO_3(OH)_3$.
Improving lakes and soils involves neutralizing the H_2SO_4 from acid rain with a calcium-containing base such as $CaCO_3$, $CaMg(CO_3)_2$, or $Ca(OH)_2$.
Part of **baking bread** is producing CO_2 from the reaction of baking soda with the acid water.
In each case, you can tell by the acid involved.

Examples:

1. $HCl + NaHCO_3 \rightarrow NaCl + CO_2 + H_2O$ STOMACH ACIDITY.
HCl is the acid. (In bread making, water is the acid.)
2. $CaCO_3 + H_2SO_4 \rightarrow CaSO_4 + CO_2 + H_2O$ LAKES AND SOILS.
H_2SO_4 is the acid. (In the stomach, HCl is the acid.)

Drill:

1. Label the following acid-base reactions as being involved in Reducing Excess Stomach **A**cidity, Improving **L**akes and Soils, or Baking **B**read:

a) $H_2O + NaHCO_3 \rightarrow NaOH + CO_2 + H_2O$ A L B

b) $HCl + CaCO_3 \rightarrow CaCl_2 + CO_2 + H_2O$ A L B

c) $Mg(OH)_2 + 2\ HCl \rightarrow MgCl_2 + 2\ H_2O$ A L B

d) $Ca(OH)_2 + H_2SO_4 \rightarrow CaSO_4 + 2\ H_2O$ A L B

e) $AlNaCO_3(OH)_2 + 4\ HCl \rightarrow$
$CO_2 + 3\ H_2O + AlCl_3 + NaCl$ A L B

Objective 4a: Classify the reactants in a given chemical reaction as being oxidized, reduced, or neither of the two.

Oxidation involves the loss of electrons, the gain of oxygen, an increase in positive electrical charge, or the loss of hydrogen.

Reduction involves the gain of electrons, the loss of oxygen, a decrease in positive electrical charge, or the gain of hydrogen.

Mnemonic: picture Leo the Lion growling and try to forget that "LEO says GER!" **L**oss of **E**lectrons is **O**xidation; **G**ain of **E**lectrons is **R**eduction.

Examples:

1. $2\ H_2 + O_2 \text{ --> } 2\ H_2O$. The H_2 reactants gain oxygen; thus, they're OXIDIZED. The O_2 must therefore be REDUCED.
2. $Zn + 2\ H^+ \text{ --> } Zn^{2+} + H_2$. Zn is OXIDIZED because it increases from no charge to 2+. H^+ is REDUCED because its charge decreases from 1+ to 0.
3. $2\ Al + 3\ Cl_2 \text{ --> } 2\ AlCl_3$. Al is OXIDIZED because it goes from being alone with all its electrons to being with more electronegative Cl and having to share. LEO; loss of electrons is oxidation. Cl_2 is REDUCED because it starts with a 50-50 share of electrons in its elemental state, but gains a lop-sided share when it combines with Al. GER; gain of electrons is reduction

Drill:

1. Indicate whether the underlined reactant is Oxidized or Reduced.

a)	$3\ \underline{Fe} + C \text{ --> } Fe_3C$	O	R
b)	$\underline{SiO_2} + 2\ Mg \text{ --> } Si + 2\ MgO$	O	R
c)	$PbS + \underline{Fe} \text{ --> } Pb + FeS$	O	R
d)	$\underline{Cu^{2+}} + Fe \text{ --> } Cu + Fe^{2+}$	O	R
e)	$4\ Ag + \underline{O_2} \text{ --> } 2\ Ag_2O$	O	R
f)	$WO_3 + 3\ \underline{H_2} \text{ --> } W + 3\ H_2O$	O	R

Objective 4b: Classify the reactants in a given chemical reaction as oxidizing agents, reducing agents, or neither of the two.

Whenever something is oxidized in a reaction it is called a **reducing agent** because by its oxidation it caused another substance to be reduced.

Likewise, whatever is reduced is called an **oxidizing agent.**

Examples:

1. In Example 1 of Objective 4a, H_2 is oxidized so it is the REDUCING AGENT. O_2 is reduced so it is the OXIDIZING AGENT. Indeed, oxygen is the most common oxidizing agent.

2. In Example 2, Zn is the REDUCING AGENT and H^+ is the OXIDIZING AGENT.
3. In Example 3, Al is the REDUCING AGENT and Cl_2 is the OXIDIZING AGENT.

Drill:

1. For each underlined reactant in Question 1 of Objective 4a, specify whether it is an Oxidizing or a Reducing agent.

Objective 5: Identify oxidation-reduction reactions involved in bleaching and stain removal, water purification, photography, and batteries.

Stain removal, or more aptly, **bleaching**, is the decolorization of stains with oxidizing agents such as OCl^-, H_2O_2, and $KMnO_4$ or with reducing agents such as $H_2C_2O_4$ and $Na_2S_2O_3$.

Water purification, for drinking or for swimming, involves the killing of germs by the oxidation of chemicals vital to these organisms. The "chlorine" used is really the oxidizing agent OCl^-

Photography takes advantage of light's ability to reduce Ag^+ to Ag on photographic film and paper.

Batteries produce electricity by requiring the electrons transferred in oxidation-reduction reactions to travel through a circuit. The reactions are usually the transfer of electrons between two different metals.

Examples:

1. $Zn + Cu^{2+} \rightarrow Zn^{2+} + Cu$. A reaction used in a BATTERY (two metals reacting). In fact, this is the reaction that was used in the "Daniell cell," one of the earliest batteries known.
2. $4\ Fe_2O_3$(rust) $+ Na_2S_2O_3 + 7\ H_2SO_4 \rightarrow 8\ FeSO_4 + Na_2SO_4 + 7\ H_2O$. A BLEACHING reaction. Rust is changed to colorless $FeSO_4$ by the oxidizing agent $Na_2S_2O_3$.

Drill:

1. Tell whether the following reactions are used in **S**tain removal or bleaching, purifying **W**ater, **P**hotography, or **B**atteries:

a) $2\ AgBr + \text{light} \rightarrow 2\ Ag + Br_2$ S W P B

b) $Ag_2O + H_2O + Zn \rightarrow 2\ Ag + Zn(OH)_2$ S W P B

c) live germ + OCl^- --> dead germ + Cl^- S W P B

d) stain + OCl^- --> colorless + Cl^- S W P B

e) $Pb + PbO_2 + 2\ H_2SO_4 \rightarrow 2\ PbSO_4 + 2\ H_2O$ S W P B

Self-Test

1. Could the reaction $Ni^{2+} + Cd \rightarrow Ni + Cd^{2+}$ be used to produce electricity in a battery?
2. The reaction $CaCO_3 + 2\ HCl \rightarrow CO_2 + H_2O + CaCl_2$ has application in
   a) reducing stomach acidity b) improving lakes or soils
   c) baking bread d) all of these
3. Is a solution with pH = 4.4 acidic, alkaline, or neutral?
4. Household bleach contains ClO^- ions which react with stains to become Cl^-. Is bleach an oxidizing agent or a reducing agent?
5. In the reaction $HS^- + OH^- \rightarrow S^{2-} + H_2O$, the acid is
   a) HS^- b) OH^- c) S^{2-} d) H_2O
6. The oxidizing agent H_2O_2, hydrogen peroxide, is used in
   a) bleaching b) water purification
   c) photography d) none to these
7. Identify the acid:
   a) solution containing OH^- b) turns litmus red
   c) tastes bitter d) $AlNa(OH)_4$
8. Flashlight batteries contain zinc metal as a source of electrons. True or false?
9. Do automobile batteries operate by this reaction?
   $$Pb + PbO_2 + 2\ H_2SO_4 \rightarrow 2\ PbSO_4 + 2\ H_2O$$
10. In the reaction $NO_2 + Mg \rightarrow MgO + NO$, the oxidizing agent is
    a) NO_2 b) Mg c) MgO d) NO
11. Which of the following would not be a good antacid for your stomach?
    a) $NaHCO_3$ b) $Al(OH)_3$ c) $CaCO_3$ d) NaOH
12. $2\ Ag^+ + Br^- \rightarrow 2\ Ag + Br_2$ is an oxidation-reduction reaction commonly used in photography. True or false?
13. Write the reaction between H^+ and OH^-.
14. Which is more acidic, a solution with pH = 8 or a solution with pH = 2?
15. What substance is reduced in the reaction $Zn + HgO \rightarrow Hg + ZnO$?
16. Which of the following is used in water purification?
    a) Cl_2 b) HCl c) ClO^- d) $HClO_4$
17. What would you expect to find in fruit punch that tastes tart and tangy, acids or bases?

Answers

Objective 1a:

1. a) A; b) B; c) B; d) B; e) B; f) A; g) A

Objective 1b:

1. The acids are: a) HCl; b) H_2SO_4; c) H_2O; d) $HC_2H_3O_2$; e) HF; f) HSO_4^-

Objective 2:

1. a) A; b) A; c) B; d) N; e) A; f) B
2. a) 1.1; b) 10; c) 7.0; d) 6.35; e) 1; f) 2

Objective 3:

1. a) B; b) A; c) A; d) L; e) A

Objective 4a:

1. a) O; b) R; c) O; d) R; e) R; f) O

Objective 4b:

1. a) R; b) O; c) R; d) O; e) O; f) R

Objective 5:

1. a) P; b) B; c) W; d) S; e) B

Self-Test:

1) yes; 2) a; 3) acidic; 4) oxidizing agent; 5) a; 6) a; 7) b; 8) true; 9) yes; 10) a; 11) d; 12) true; 13) $H^+ + OH^- \rightarrow H_2O$; 14) pH = 2; 15) HgO; 16) c; 17) acids

Evaluation:

If you missed more than one question on the self-test in any of the following groups, you need to review the section indicated:

Question Groups	Section
3, 5, 7, 14, 17	6.1
2, 11, 12	6.2
4, 10, 15	6.3
6, 12, 16	6.4
1, 8, 9	6.5

Chapter 7

Energy and Speed of Chemical Reactions: Influencing Chemical Changes

---

Outline

I. The First Law of Energy: You Can't Get Something for Nothing
    A. Energy Changes in Natural Processes
        1. Forms of energy
        2. Changes from one form to another
    B. The First Law of Energy
        1. Conservation of energy
        2. System and surroundings
    C. Spontaneous processes
    D. The Tendency Toward Minimum Potential Energy
        1. Exothermic and endothermic reactions
        2. Minimum energy hypothesis
            a. Tendency toward minimum energy
            b. All spontaneous reactions should be exothermic
        3. Spontaneous endothermic reactions
II. The Second Law of Energy: You Can't Even Break Even
    A. The Second Energy Law and Energy Quality
        1. High quality (more concentrated) and low quality (less concentrated) energy
        2. When energy changes form, some energy is always degraded in quality
        3. Efficiency of various energy conversions
    B. The Second Law and Increasing Disorder
        1. Heat energy flows from hot to cold
        2. Any process spontaneously increases disorder
            a. Things fall into disrepair
            b. Vegetation does not naturally grow in straight rows
            c. Dye colors, odors, and smoke spread
            d. Solution to pollution is often dilution
        3. Entropy
            a. The scientific measure of disorder
            b. Increases from solid to liquid to gas
III. Reaction Rate: How Can We Make a Reaction Go Faster or Slower?
    A. What is Reaction Rate?
        1. Speed at which reactants are converted to products
        2. Measured by monitoring change in concentrations
    B. What Determines Reaction Rate?
        1. Fast and slow reactions
        2. Collision frequency
        3. Collision energy and activation energy
        4. Collision orientation and geometry

C. How Can We Alter the Rate?
   1. Change concentration of reactants
   2. Change temperature
   3. Change degree of subdivision of reacting particles
   4. Add catalyst or inhibitor or enzyme

IV. Dynamic Chemical Equilibrium and LeChatelier's Principle: How Far Will a Reaction Go?

A. Dynamic Equilibrium
   1. Products of reactions can revert back to reactants
   2. Equilibrium occurs when rate of forward reaction equals that of the reverse
   3. Not a static process

B. Making a Reaction Go Farther: LeChatelier's Principle
   1. If an equilibrium is subjected to a stress, the system will shift to relieve the stress
   2. Adding more reactants favors the forward reaction; adding more products favors the reverse
   3. Heating favors the endothermic direction; cooling favors the exothermic

C. Chemical Fraud
   1. A swindle involving the making of hydrogen fuel from water
   2. The victim did not know the criteria for a practical reaction
      a. Energy questions
      b. Rate questions
      c. Yield questions

Objectives

After you read and study the chapter [and the sections in brackets], you should be able to:

[M] 1. Give statements of the first and second laws of thermodynamics. [Sections 7.1 and 7.2; Questions 1, 7, 8, 9, 10]

[C] 2. Distinguish between and identify examples of:
   a. Kinetic and potential energy. [Section 7.1; Question 2]
   b. Systems and surroundings. [Section 7.1; Question 6]
   c. Exothermic and endothermic reactions. [Section 7.1; Question 3]

[R] 3. Identify which of two given systems has the higher entropy, based on differences in concentration, temperature, or physical state. [Section 7.2; Question 5]

[C] 4. Identify processes that illustrate the tendency toward (a) increasing disorder or (b) minimum energy. [Sections 7.1 and 7.2; Question 4]

[R] 5. Give the effect on the rate of a reaction to a given (a) change in the concentration of a reactant, (b) change in temperature, (c) change in subdivision of a reactant, or (d) addition of a catalyst or inhibitor. Section 7.3; Questions 11, 12, 13, 14, 15, 16, 17, 18]

[R] 6. Give the effect on a reaction in equilibrium due to a change in (a) the concentration of a reactant or product, or (b) the temperature (given which reaction direction is exothermic). [Section 7.4; Questions 19, 20, 21]

Practice

Objective 1: Give statements of the first and second laws of thermodynamics.

Drill:
Cut out the flash cards, and memorize the material on them.

Objective 2a: Distinguish between and identify examples of kinetic and potential energy.

Kinetic energy is the energy of movement and includes all mechanical, electrical, thermal, and light energy.
Potential energy is the energy of position and becomes kinetic energy only after it is unleashed. It is always constrained in some way.

Examples:

1. The movement of a falling rock is KINETIC ENERGY; the position of a rock at the edge of a cliff is POTENTIAL ENERGY (the rock would roll or fall downward if not perched on the cliff); the position of a rock on flat ground is NEITHER (the rock has no place to fall or roll).
2. A compressed spring has POTENTIAL ENERGY which turns into KINETIC ENERGY when it is sprung. Then the spring has NEITHER; its energy has been transferred to some other system.
3. The bonds in molecules of nitroglycerine have chemical POTENTIAL ENERGY. This energy is only barely constrained. The slightest jolt will turn it into the KINETIC ENERGY of an explosion.

Drill:

1. What kind of energy does each of the following possess: Kinetic, Potential, or Neither of the two?
   a) water flowing through a garden hose **K P N**
   b) water in the hose with the faucet on full but the nozzle shut tight **K P N**
   c) water left in the hose after the faucet is turned off **K P N**
   d) the chemical bonds in gasoline **K P N**
   e) the illumination from a candle flame **K P N**
   f) liquid nitrogen boiling at -196°C **K P N**
   g) the shade from a tree on a sunny day **K P N**
   h) litmus turning from red to blue in an acid-base reaction **K P N**

Flash Cards for Objective 1

In all chemical and physical changes, energy is neither created nor destroyed but merely transformed from one to another	In any process, the total energy of the system plus its surroundings or environment remains constant
You can't get something for nothing	You can't even break even
Any system plus its surroundings tends spontaneously toward increasing entropy or disorder	In any conversion of energy to useful work, some energy is always degraded to a less useful form
Heat energy flows spontaneously from hot to cold	

First Law of
Thermodynamics

First Law of
Thermodynamics

Second Law of
Thermodynamics

First Law of
Thermodynamics

Second Law of
Thermodynamics

Second Law of
Thermodynamics

Second Law of
Thermodynamics

Objective 2b: Distinguish between and identify examples of systems and surroundings.

A **system** is any part of the universe you're interested in at the moment--from your point of view. You are free to choose which part, big or small. For purposes of chemistry, however, the system is usually the atoms, ions or molecules involved in a particular reaction.

The **surroundings** are all the rest of the universe; that is, they are literally everything that isn't the system, no matter how near or far away. In most chemical situations, you and your body are part of the surroundings.

Examples:

1. In measuring the statewide birth control rate, a statistician would use everyone in the state as the SYSTEM. All other people and things would be the SURROUNDINGS.
2. To find the heat of combustion of natural gas, a chemist would use the reaction

   $$CH_4 + 2\ O_2 \longrightarrow CO_2 + H_2O$$

   as the SYSTEM; that is, the molecules involved. The reaction vessel, the laboratory, all the people in Example 1, and everything else would be the SURROUNDINGS.
3. On the other hand, if an anthropologist were interested in the effect natural-gas heat had on civilization, then, for him or her, the SYSTEM and SURROUNDINGS in Example 2 would be reversed.

Drill:

1. Identify the system in each of the following studies:
   a) finding out how fast rust forms on the fender of your car.
   b) finding out how fast iron rusts under laboratory conditions.
   c) measuring the energy efficiency of your home water heater
   d) discovering the amount of hot water used in your home
   e) ascertaining the quantity of carbon monoxide produced when gasoline is burned.

Objective 2c: Distinguish between and identify examples of exothermic and endothermic reactions.

Exothermic reactions give off heat. The surroundings (the container, the observer, the laboratory, etc.) get warmer. The molecules in the reaction lose energy.
Endothermic reactions are just the opposite. They take in or absorb heat. The surroundings get cooler, and the system gains energy.

Examples:

1. Wood burning in the fireplace is EXOTHERMIC. The room gets warm because the logs lose energy and transfer it to the room.
2. Ice melting is ENDOTHERMIC because water in the solid state requires energy to change to the liquid state. This energy comes from the environment around the ice.
3. Water boiling is also ENDOTHERMIC even though it might scald you. Energy must be added to the water to make it boil and to keep it boiling. It's not the temperature that matters; it's whether energy goes in or out of the system. And since it must go into the system (liquid water in this case), boiling is endothermic.

Drill:

1. State whether these reactions are e**X**othermic or e**N**dothermic.

a)	water evaporating from your skin	**X**	**N**
b)	butter melting	**X**	**N**
c)	charcoal glowing in a barbeque	**X**	**N**
d)	steaks broiling on the barbeque	**X**	**N**
e)	natural gas burning in a furnace	**X**	**N**

2. You are holding a beaker of water, and you pour some concentrated sulfuric acid into it. The beaker gets so hot that you can hardly continue to touch it. Was the reaction exothermic or endothermic?
3. (a) Stretch a rubber band, and quickly touch it to your tongue. It feels warm. Why? (b) Puncture a CO_2 cartridge, and let the gas escape rapidly. The cartridge becomes ice-cold. Why?

Objective 3: Identify which of two given systems has the higher entropy, based on differences in concentration, temperature, or physical state.

Rule: A substance has greater entropy or disorder as its temperature increases and as it changes state from solid to liquid to gas. A dissolved substance has more entropy when it is less concentrated.

Procedure: To choose the system with greater entropy:

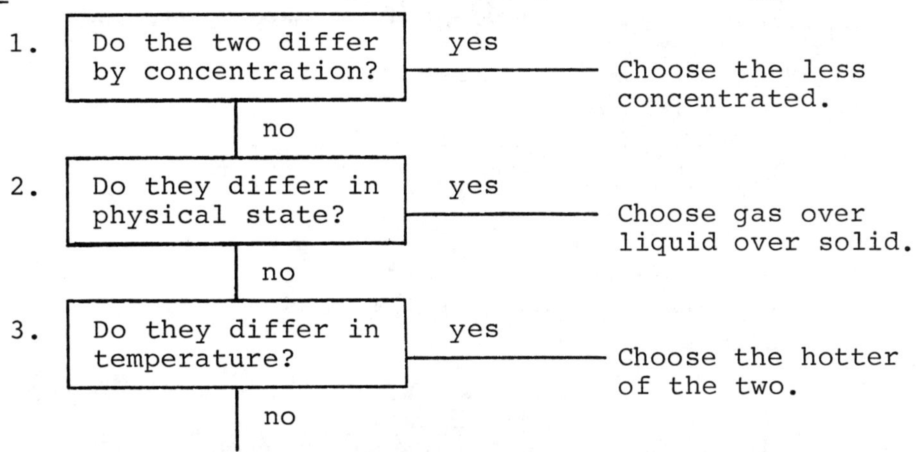

1.
Do the two differ by concentration?
yes
Choose the less concentrated.
no
2.
Do they differ in physical state?
yes
Choose gas over liquid over solid.
no
3.
Do they differ in temperature?
yes
Choose the hotter of the two.
no
Their entropies can't be distinguished by this rule

Examples:

1. Hot water compared to cool water. Question 1: no; question 2: no; question 3: yes => hot water has the greater entropy.
2. Salt in the ocean compared to salt in the Great Salt Lake. Question 1: yes => salt is less concentrated in the ocean; thus, it has more entropy there.
3. Steam compared to ice. Question 1: no; question 2: yes => the gas (steam) has more entropy that the solid (ice).

Drill:

1. In each of the following pairs, circle the one with the greater entropy.
   a) air on top of Mt. Everest compared to air at the same temperature in International Falls, Minnesota
   b) gasoline in your tank compared to gasoline vapor in your fuel line.
   c) cold soda pop compared to warm
   d) sugar in fruit punch compared to sugar in pancake syrup
   e) CO_2 in your breath compared to dry ice (solid CO_2)
2. Place in order of increasing entropy: compressed nitrogen gas in a tank; liquid nitrogen in a thermos; nitrogen gas in the air.
3. Circle the one with lesser entropy in each of the following:
   a) a branding iron in the fire compared to a branding iron in the saddle bags.
   b) solid I_2 compared to gaseous I_2.
   c) SO_2 air pollution, 1 km from a copper smelter compared to SO_2 pollution 10 km from the smelter.

d) air in the summer compared to air in the winter.
e) anhydrous ammonia fertilizer (pure NH_3 liquid) compared to household ammonia cleanser (NH_3 dissolved in water).

Objective 4: Identify processes that illustrate the tendency toward (a) increasing disorder or (b) minimum energy.

Processes that illustrate the tendency toward **increasing disorder** are those that achieve a state of higher entropy. They include, but are not limited to, all spontaneous, endothermic processes and reactions.
Processes that illustrate the tendency toward **minimum energy** are all exothermic processes and reactions. (Many of these illustrate both tendencies.)

Examples:

1. The melting of ice at room temperature. INCREASING DISORDER (spontaneous and endothermic).
2. The burning of coal. BOTH (exothermic and achieves higher entropy).
3. The exothermic formation of solid AgCl when a solution of $AgNO_3$ is mixed with hydrochloric acid. MINIMUM ENERGY (exothermic but decreases in entropy).

Drill:

1. Tell whether the following processes illustrate a tendency toward Increasing Disorder, Minimum Energy, Both, or Neither:

a)	the sudden (and endothermic) expansion of air	D	E	B	N
b)	the explosion of dynamite	D	E	B	N
c)	the freezing of ice	D	E	B	N
d)	the endothermic formation of solid HgS from solution	D	E	B	N
e)	the formation of CO_2 gas when $CaCO_3$ is treated with acid (exothermic)	D	E	B	N
f)	dew forming on the front lawn	D	E	B	N
g)	your metabolism turning food into energy	D	E	B	N
h)	a battery discharging	D	E	B	N

Objective 5: Give the effect on the rate of a reaction to a given (a) change in the concentration of a reactant, (b) change in temperature, (c) change in subdivision of a reactant, or (d) addition of a catalyst or inhibitor.

Rule: The rate of a reaction will be increased if (a) the concentration of any reactant is increased, (b) the temperature is raised, (c) any reactant is subdivided into smaller pieces, or (d) a catalyst is added. The opposite of any of these will decrease the rate.

Procedure: To choose which condition will speed up a reaction:

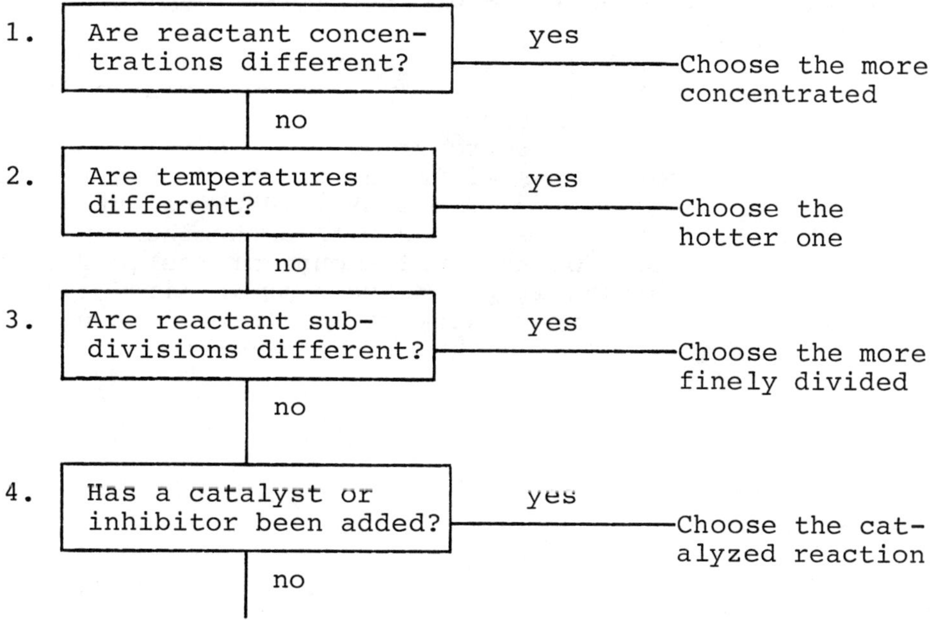
1.
Are reactant concentrations different?
yes
Choose the more concentrated
no
2.
Are temperatures different?
yes
Choose the hotter one
no
3.
Are reactant subdivisions different?
yes
Choose the more finely divided
no
4.
Has a catalyst or inhibitor been added?
yes
Choose the catalyzed reaction
no
The reaction rate is the same under both conditions.

Examples:

1. Small chunks of coal compared to large chunks burn (question 1: no; question 2: no; question 3: yes--small chunks more finely divided) faster.
2. Hydrogen peroxide decomposes in the presence of iron, a catalyst, (questions 1, 2, and 3: no; question 4: yes) faster.
3. Boiling potatoes in a open pan compared to boiling them in a pressure cooker where it's hotter is (question 1: no; question 2: yes) slower.
4. The explosion of pure nitroglycerine compared to dynamite (nitroglycerine mixed with kieselguhr) is (question 1: yes--pure nitro is more concentrated) more rapid and thus more violent.

Drill:

1. Choose which of the two reactions is faster:
   a) igniting fireworks with more oxidizer compared to less
   b) dissolving granulated sugar compared to powdered sugar
   c) decomposing auto exhaust air pollutants in a catalytic converter compared to in an ordinary muffler
   d) getting drunk on beer or vodka
   e) paint drying in an oven compared to in a refrigerator
2. Which of the two reactions is slower?
   a) hot dogs spoiling at the Green Bay Packers' stadium or at the Miami Dophins' stadium
   b) cooling a hot tub with a wheelbarrow load of ice cubes or an equal amount of ice in one large block
   c) burning in pure oxygen or in air
   d) soda pop growing mold with or without an inhibitor

Objective 6: Give the effect on a reaction in equilibrium due to a change in (a) the concentration of a reactant or product, or (b) the temperature (given which reaction direction is exothermic).

Rule: Adding reactants favors the forward reaction; adding products favors the reverse. Removing reactants favors the reverse reaction; removing products favors the forward. Increasing the temperature favors the endothermic direction (in which heat is a reactant); decreasing the temperature favors the exothermic (in which heat is a product). (This rule is entirely independent of the previous reaction rate rule).

Procedure: To determine which direction an equilibrium will shift under a given stress:

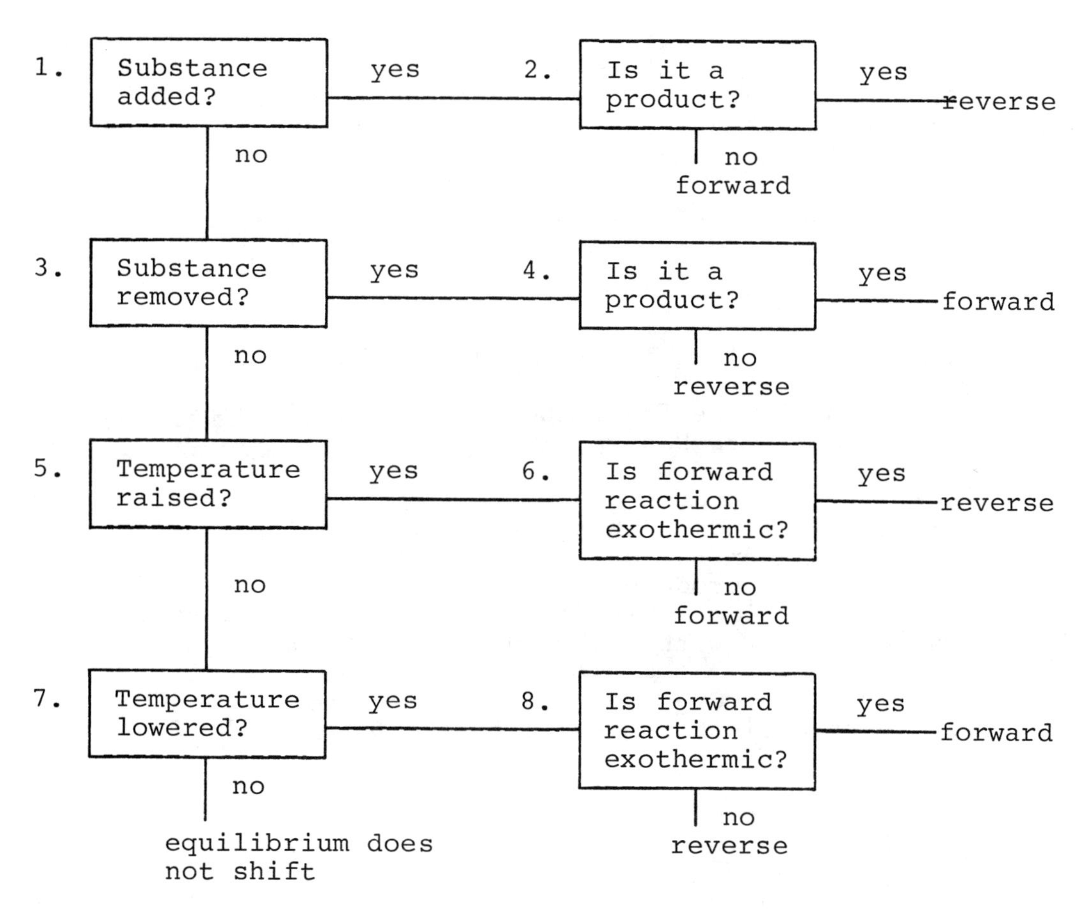

Examples:

Anhydrous ammonia fertilizer (NH_3) is made by the Haber process:

$$N_2 + 3\ H_2 \text{ --> } 2\ NH_3 + \text{heat}$$

1. When extra N_2 is added, the equilibrium shifts (question 1: yes; question 2: no) in the forward direction, or to the right.
2. If the temperature is lowered, the equilibrium shifts (question 1: no; question 3: no; question 5: no; question 7: yes; question 8: yes) forward also.
3. If the reaction vessel develops a slow leak, H_2 escapes more than anything else. This causes the equilibrium to shift (question 1: no; question 3: yes; question 4: no) in the reverse direction, or to the left.

Drill:

1. For the reaction:

$$2\ NO_2 \quad \text{-->} \quad N_2O_4 \quad + \quad \text{heat}$$

Predict whether the equilibrium will shift in the forward (-->) direction or in the reverse (<--) when the following changes are made:

a) the temperature is raised	-->	<--
b) NO_2 is added	-->	<--
c) N_2O_4 is removed	-->	<--

2. Ozone (O_3) is made from oxygen (O_2):

$$3\ O_2 \quad + \quad \text{heat} \quad \text{-->} \quad 2\ O_3$$

Which conditions favor the Production of ozone, and which favor its Decomposition?

a) low temperatures	**P**	**D**
b) high altitudes (where there is less O_2)	**P**	**D**
c) high concentrations of O_3	**P**	**D**

3. Steel makers produce a mixture of steam (H_2O) and carbon monoxide (CO), called water gas, by reacting hydrogen (H_2) with carbon dioxide (CO_2):

$$H_2 \quad + \quad CO_2 \quad + \quad \text{heat} \quad \text{-->} \quad H_2O \quad + \quad CO$$

Will the equilibrium shift to the Left or the Right when:

a) CO is removed	**L**	**R**
b) the blast furnace is turned off	**L**	**R**
c) H_2 is added	**L**	**R**
d) H_2O is added	**L**	**R**

Self-Test

---

1. "You can't get something for nothing." Is this the first or second law of thermodynamics?
2. When jello gels, does it increase or decrease in disorder?
3. In a closed container, the following reaction among gases proceeds:

$$2\ NO + 2\ H_2 \text{ --> } N_2 + 2\ H_2O + \text{heat}$$

For each of the following, tell whether the condition will speed Up, slow Down, or Not affect the rate of the reaction:

a) cooling the container U D N
b) adding H_2 U D N
c) adding H_2O U D N
d) adding NO U D N
e) adding a catalyst U D N

When the reaction reached equilibrium, did the following changes shift the equilibrium to the Left, the Right, or Neither direction?

f) cooling the container L R N
g) adding H_2 L R N
h) adding H_2O L R N
i) adding NO L R N
j) adding a catalyst L R N

4. Is the reaction in Question 3 exothermic or endothermic?
5. When does a dissolved substance have greater entropy, in a dilute or concentrated solution?
6. Does the water flowing in a river have kinetic energy, potential energy, or both?
7. True or false, the second law of thermodynamics requires every change to increase the entropy of a system.

Answers

Objective 2a:

1. a) K; b) P; c) N; d) P; e) K; f) K; g) N; h) K

Objective 2b:

1. a) the fender; b) the atoms and molecules; c) the water heater; d) your home; e) the atoms and molecules

Objective 2c:

1. a) N; b) N; c) X; d) N; e) X
2. exothermic
3. a) stretching must be exothermic
   b) gas expansion must be endothermic

Objective 3:

1. a) air on Mt. Everest; b) gasoline vapor; c) warm soda pop; d) sugar in punch; e) CO_2 in your breath
2. Nitrogen in thermos < in tank < in air
3. a) in saddle bags; b) solid I_2; c) 1 km from smelter; d) in winter; e) fertilizer

Objective 4:

1. a) D; b) B; c) E; d) N; e) B; f) E; g) B; h) B

Objective 5:

1. a) more oxidizer; b) powdered sugar; c) in catalytic converter; d) vodka; e) in an oven
2. a) at Green Bay (cooler); b) block; c) in air; d) with inhibitor

Objective 6:

1. a) <--; b) -->; c) -->
2. a) D; b) D; c) D
3. a) R; b) L; c) R; d) L

Self-Test:

1) first law; 2) decrease; 3a) D; b) U; c) N; d) U; e) U; f) R; g) R; h) L; i) R; j) N; 4) exothermic; 5) dilute; 6) both; 7) false

Evaluation:

If you missed more than one question on the self-test in any of the following groups, you need to review the section indicated:

Question Groups	Section
1, 4, 6	7.1
2, 5, 7	7.2
3 a-e	7.3
3 f-j	7.4

Chapter 8

Organic Chemistry: Some Important Carbon Compounds

Outline

I. Carbon: A Unique Element
- A. Why Does Carbon Form So Many Compounds?
- B. Forms of the Element Carbon

II. Hydrocarbons: Chains and Rings
- A. Saturated Hydrocarbons: The Simple Alkanes
- B. Structural Isomers
- C. Unsaturated Hydrocarbons
- D. Unsaturated Hydrocarbon Rings: Aromatic Hydrocarbons

III. Fossil Fuels and Petrochemicals: From Oil to Plastics
- A. Importance of Fossil Fuels
- B. Coal
- C. Natural Gas
- D. Petroleum
- E. Petroleum Refining
- F. Gasoline Production

IV. Functional Groups: Where the Action Is

V. Organic Halides: From Refrigerants to Pesticides

VI. Oxygen-containing Groups: From Cocktails to Flavors
- A. Alcohols and Ethers
- B. Aldehydes and Ketones
- C. Organic Acids and Esters

VII. Nitrogen-containing Groups: From Drugs to Dynamite
- A. Amines
- B. Amides
- C. Amino acids

Objectives

After you read and study the chapter [and the sections in brackets], you should be able to:

[C] 1. Explain why carbon forms so many more compounds than other elements. [Section 8.1; Question 1]

[C] 2. Recognize structural isomers, and how to represent molecules by condensed structural formulas. [Section 8.2; Questions 4, 5]

[C] 3. Distinguish between saturated, unsaturated, and aromatic hydrocarbons and give examples of each. [Section 8.2; Questions 2, 3]

[C] 4. Name and identify some important hydrocarbons and their uses, and explain how they are supplied by fossil fuels. [Section 8.3; Questions 6, 7, 11]

[C] 5. Identify the following functional groups: organic halides, alcohols, ethers, aldehydes, ketones, carboxylic acids, esters, and amines. [Sections 8.4, 8.5, 8.6, 8.7; Questions 8, 9, 10]

[C] 6. Explain some common uses of organic halides, alcohols, ethers, aldehydes, ketones, carboxylic acids, esters, and amines. [Sections 8.4, 8.5, 8.6, 8.7; Questions 12, 13]

Practice

Objective 1: Explain why carbon forms so many more compounds than other elements.

Unlike atoms of other elements, carbon atoms can form stable covalent bonds with other like (carbon) atoms to form large and complex molecular structures. This helps give carbon its unique ability to form a remarkable variety of molecules.

Examples:

1. Carbon can form long chains containing hundreds and even thousands of carbon atoms.
2. Carbon can form branched chains and ring structures, too.
3. Carbon can join to other carbon atoms with single, double, or triple covalent bonds. Carbon also forms covalent bonds with other nonmetals, especially hydrogen, oxygen, nitrogen, fluorine, chlorine, and bromine.

Drill:

1. Which of the following elements would not normally form a covalent bond with carbon: a) S, b) O, c) Na, d) C, e) Br
2. From the periodic chart, you would predict that the element that would most closely resemble carbon's properties, and thus also might form a wide variety of compounds, is a) B, b) N, c) Ca, d) Si, e) O
3. Which of the following does carbon not normally form: a) ring structures, b) covalent compounds, c) very large molecules, d) chains of carbon atoms, e) none of the above

Objective 2: Recognize structural isomers, and how to represent molecules by condensed structural formulas.

Structural isomers are compounds with the same composition (molecular formula) that differ in which atoms are bonded to each other. They also have different physical and chemical properties.

Condensed structural formulas are a time- and space-saving way to represent structural formulas. With straight-chain and branched compounds, horizontal single bonds are omitted, and all of the atoms bonded to a carbon atom are written together right after the carbon atom. With ring compounds, a geometric figure is used with the same number of sides as atoms in the ring. Only noncarbon atoms are shown in the corners of the ring, and only nonhydrogen atoms are shown attached to ring atoms. Benzene may be represented as

or

Examples:

1.
```
  H H           H   H
H-C-C-O-H and H-C-O-C-H
  H H           H   H
```
both have the same number of carbon, hydrogen, and oxygen atoms; both have the molecular formula C_2H_6. But they differ in which atoms are bonded to each other, so they are structural isomers with different physical and chemical properties. As you will see below (in Objective 5), the first compound is an alcohol and the second is an ether.

2. Listed below are several substances written in both their extended and condensed structural formulas:

Extended formula / **Condensed structural formula**

```
   H H H
Cl-C-C-C-O-H
   H H H
```
$CH_2ClCH_2CH_2OH$

```
    H
  H-C-H
  H | H O H Br
H-C-C-C-C-C-C-H
  H H H   H H
```
```
   CH3
   |      O
CH3CHCH2CCH2CH2Br
```

```
        Cl
H       C       H
    C       C
    C       C
H       C       H
        Cl
```
```
 Cl
(ring)
 Cl
```

Drill:

1. Which of the following pairs are structural isomers:

```
      H H   H             H H H H
a)  H-C-C-O-C-H    and  H-C-C-C-C-H
      H H   H             H H H H

      H H H O             H O H H
b)  H-C-C-C-C-H    and  H-C-C-C-C-H
      H H H               H   H H

      H H H H             H H H H H
c)  H-C-C-C-C-H    and  H-C-C-C-C-C-H
      H H H H             H H H H H

      H H H H             H H H
d)  H-C-C-C-C-H    and  H-C-C-C-H
      H H H H             H | H
                          H-C-H
                            H
```

2. Write condensed structural formulas for the following:

   a) each of the compounds in Question 1 above, parts b and d

```
b)   H  H                 c)      H   O—H
      \ /                          \ /
       C     H               H      C      H
      / \    |                \   /   \   /
  H—C———C — C—H              H—C       C—H
    /    \   |               H—C       C—H
   H      H  H                  /  \   /   \
                               H     C      H
                                    / \
                                   H   H
```

Objective 3: Distinguish between saturated, unsaturated, and aromatic hydrocarbons and give examples of each.

Saturated hydrocarbons have only single bonds between carbon atoms. They include the alkanes.

Unsaturated hydrocarbons have one or more double bonds or triple bonds between adjacent carbon atoms; these are classified as alkenes and alkynes, respectively.

Aromatic hydrocarbons contain a benzene or benzenelike ring. Benzene is a cyclic structure with the formula C_6H_6 that has a donut-shaped space above and below the plane of the ring occupied by six additional electrons that are shared equally by the single-bonded carbons in the ring.

Examples:

1. C_3H_8, propane, and C_8H_{18}, octane, are saturated hydrocarbons that are members of the alkane family.

2. Ethene (ethylene) is the simplest alkene. It is an unsaturated compound with the structural formula

```
  H H
  | |
H-C=C-H
```
or $CH_2=CH_2$.

3. Ethyne (acetylene) is the simplest alkyne. It is an unsaturated compound with the structural formula H-C≡C-H or CH≡CH.

4. Aromatic compounds such as benzene and toluene have the following structural formulas, respectively:

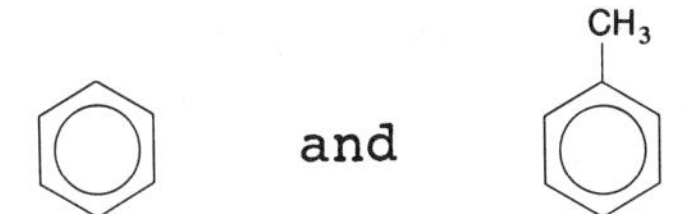

Drill:

1. Classify the following compounds as saturated, unsaturated, or aromatic hydrocarbons:

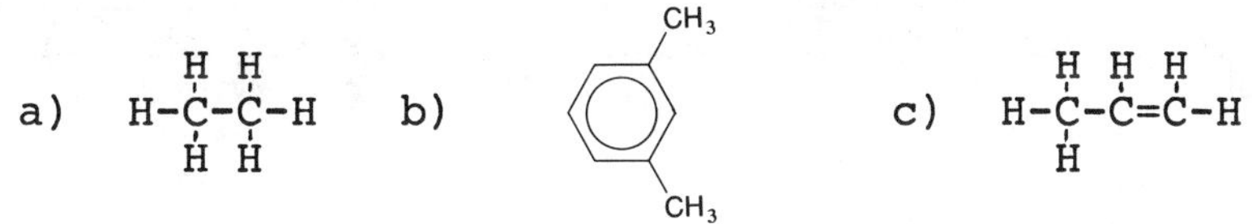

Objective 4: Name and identify some important hydrocarbons and their uses, and explain how they are supplied by fossil fuels.

The linear alkanes containing 1 to 10 carbon atoms are named, respectively, methane, ethane, propane, butane, pentane, hexane, heptane, octane, nonane, and decane. All are used as fuels, and some (such as hexane) are solvents for nonpolar materials.

Ethene (ethylene), the simplest alkene, is used to ripen fruits and vegetables and to make the plastic polyethylene (Section 15.2). Ethyne (acetylene), the simplest alkyne, can produce a hot flame for welding.

The most important aromatic compound is benzene.

The fossil fuels include coal, natural gas, and petroleum. Coal is mostly carbon, and it can be processed to form aromatic and small linear organic compounds. Natural gas is mostly methane, with small amounts of ethane, propane, and longer-chain hydrocarbons. Petroleum contains a wide variety of hydrocarbons containing from 1 to more than 20 carbon atoms. Fossil fuels are the main source of organic chemicals used commercially.

Examples:

1. Propane, C_3H_8, is a light fuel used in camp stoves. It occurs naturally in petroleum and natural gas.
2. Heating coal in the absence of air produces coal tar, which is a rich source of aromatic compounds.
3. Gasoline typically contains hydrocarbons in the range of 5 to 12 carbon atoms. One example is an isomer of octane, C_8H_{18}.

Drill:

1. List the three major fossil fuels.
2. An alkane containing 4 carbon atoms is named a) ethane, b) butane, c) methane, d) pentane, e) octane
3. Acetylene used for welding torches is an a) alkane, b) alkene, c) alkyne, d) aromatic compound, e) none of the above
4. Natural gas contains alkanes with mostly how many carbon atoms: a) 1, b) 2, c) 3, d) 4, e) more than 4

Objective 5: Identify the following functional groups: organic halides, alcohols, ethers, aldehydes, ketones, carboxylic acids, esters, and amines.

We will use *R* to represent a general hydrocarbon group of one or more carbon atoms and their bonded hydrogen atoms. The rest of the molecule is the functional group, the region most responsible for the physical and chemical properties of a substance. Listed below (and in Table 8.9 in the text) are the general formulas for various functional groups:

organic halides	R-X (X can be F, Cl, Br, or I)
alcohols	R-O-H
ethers	R-O-R'
aldehydes	$R\text{-}\overset{\text{O}}{\overset{\|}{\text{C}}}\text{-H}$
ketones	$R\text{-}\overset{\text{O}}{\overset{\|}{\text{C}}}\text{-}R'$
carboxylic acids	$R\text{-}\overset{\text{O}}{\overset{\|}{\text{C}}}\text{-O-H}$
esters	$R\text{-}\overset{\text{O}}{\overset{\|}{\text{C}}}\text{-O-}R'$
amines	$R\text{-}\underset{\text{H}}{\underset{\|}{\text{N}}}\text{-H}$

Examples:

1.
```
    H H
    | |
H-C-C-O-H is an alcohol.
    | |
    H H
```

2.
```
    H H O H H
    | | ‖ | |
H-C-C-C-C-C-H is a ketone.
    | |   | |
    H H   H H
```

```
       H H O    H H
3.   H-C-C-C-O-C-C-H is an ester.
       H H      H H
```

Drill:

1. Identify the functional group present in the following:

```
      H H   H              H H H
a)  H-C-C-O-C-H       b) H-C-C-N-H
      H H   H              H H

       H H                 H O
c)  Cl-C-C-H          d) H-C-C-H
       H H                 H
```

Objective 6: Explain some common uses of organic halides, alcohols, ethers, aldehydes, ketones, carboxylic acids, esters, and amines.

Organic halides have been used as anesthetics, insecticides, refrigerant coolants, aerosol materials, and temporary blood substitutes.
Alcohols are used for drinking, fuel, antifreeze, antiseptics, and moisturizing agents.
Ethers are used as anesthetics and nonpolar solvents.
Aldehydes are used as preservatives and in insulation. Certain aldehydes and ketones are used as flavoring agents and in perfumes. Ketones also are solvents to remove varnish, paint, and fingernail polish.
Carboxylic acids are major ingredients in vinegar and ant stings. Esters are used as flavoring agents and in perfumes. Fats and oils are esters. Synthetic polyesters, such as Dacron, contain many ester functional groups.
Amines include many drugs (whose names end in -ine), some vitamins, and amino acids, which form proteins in our bodies.

Examples:

1. Vinegar is a 4-5% solution of acetic acid, $CH_3\overset{O}{\overset{\|}{C}}-OH$.
2. $CHCl_3$ (chloroform) and $CH_3CH_2-O-CH_2CH_3$ (diethyl ether) are an organic halide and ether, respectively, that have been used as anesthetics.
3. Caffeine, Novocaine, amphetamine, and thiamine (a B vitamin) are amines, as their -ine suffix indicates.
4. Embalming fluids and solutions to fix and preserve biological specimens for study often contain the simplest aldehyde, formaldehyde, which has the structure $H-\overset{O}{\overset{\|}{C}}-H$.

Drill:

1. Classify each of the following as an organic halide, alcohol, ether, aldehyde, ketone, carboxylic acid, ester, or amine:

   a) ________ main component in antifreeze
   b) ________ used as a coolant liquid in refrigerators
   c) ________ material from which proteins are made
   d) ________ solvent to remove varnish and paint
   e) ________ fragrant material used in perfumes

Self-Test

---

1. Organic chemistry is the study of compounds based on the element a) H, b) O, c) C, d) N, e) none of the above
2. Structural isomers do <u>not</u> differ from each other in terms of their a) melting points, b) structural formulas, c) solubility in water, d) molecular formulas, e) boiling points
3. The main organic material in vinegar is a(n) a) organic halide, b) carboxylic acid, c) ketone, d) ester, e) amine
4. List the functional group present in each of the following:

```
     H H            H H H                  O                H H H
a) H-C-C-O-H   b) H-C-C-C-Br   c)         ||         d) Cl-C-C-C-H
     H H            H | H            H     C     H          H | H
                    H-C-H                 / \               N-H
                      H           H -  C     C  - H          H
                                  H  /         \  H
                                    C --------- C
                                  H/ |         | \H
                                     H         H
```

5. Write condensed structural formulas for the compounds in Question 4.
6. List the three major types of fossil fuels.
7. Butane is a ________ (saturated or unsaturated) hydrocarbon in the ________ (alkane, alkene, alkyne, or aromatic) family that has __ carbon atoms.
8. Aromatic compounds typically contain a) alkane chain, b) chlorine, c) benzene ring, d) carbon-to-carbon triple bonds, e) ketones
9. Carbon is unique among elements in terms of its a) ability to form ionic bonds, b) electronegativity, c) ability to form long chains with like atoms, d) ability to conduct electricity, e) ability to form four covalent bonds
10. A compound with the formula C_4H_6 could be a) saturated, b) unsaturated, c) an alcohol, d) aromatic, e) an amine
11. Substances used as anesthetics include a) organic halides, b) esters, c) carboxylic acids, d) ethers, e) alkynes
12. Natural gas contains mostly ________, which has the molecular formula ________ and is classified as a(n) ________.
13. In addition to carbon and hydrogen, amines contain the element a) N, b) S, c) F, d) O, e) none of the above

14. The fossil fuels are the raw materials from which a wide variety of organic products are made. These basic chemical products are collectively called ____________.

Answers

Objective 1:

1. c
2. d
3. e

Objective 2:

1. b and d both contain a pair of structural isomers
2. a) $CH_3CH_2CH_2\overset{O}{\overset{\|}{C}}H$ and $CH_3\overset{O}{\overset{\|}{C}}CH_2CH_3$ for part b

   $CH_3CH_2CH_2CH_3$ and $CH_3\underset{CH_3}{\underset{|}{CH}}CH_3$ for part d

   b) CH_3

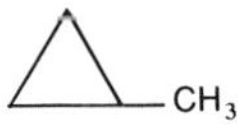

   c) OH

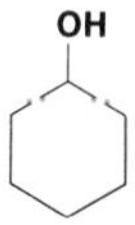

Objective 3:

1. a) saturated; b) aromatic; c) unsaturated (alkene)

Objective 4:

1. coal, natural gas, petroleum
2. b
3. c
4. a

Objective 5:

1. a) ether; b) amine; c) organic halide; d) aldehyde

Objective 6:

1. a) alcohol; b) organic halide; c) amine and carboxylic acid functional groups; d) ketone; e) esters (also certain aldehydes and ketones)

Self-Test:

1) c; 2) d; 3) b; 4a) alcohol; 4b) organic halide; 4c) ketone; 4d) amine and organic halide; 5a) CH_3CH_2OH; 5b) CH_3CHCH_2Br ; 5c) 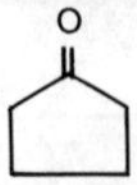 O ; 5d) $CH_2ClCHCH_3$

(5b: CH_3 branch on CH; 5d: NH_2 branch on CH)

6) natural gas, coal, petroleum; 7) saturated, alkane, 4; 8) c; 9) c; 10) b; 11) a and d; 12) methane, CH_4, alkane (or saturated hydrocarbon); 13) a; 14) petrochemicals

Evaluation:

If you missed more than one question on the self-test in any of the following groups, you need to review the section indicated:

Question Groups	Section
1, 9	8.1
2, 5, 7, 8, 10	8.2
6, 12, 14	8.3
4a-4d	8.4
4b, 4d, 11	8.5
3, 4a, 4c, 11	8.6
4d, 13	8.7

Chapter 9

Biochemistry: Some Important Molecules of Life

Outline

I. Biochemistry: Where There's Life, There's Carbon
   A. Vital Molecules
   B. Biopolymers
II. Carbohydrates: Sugar and Rice and Everything Nice
   A. Fuel for Your Body
   B. Simple Sugars, or Monosaccharides
      1. Pentoses (ribose, deoxyribose)
      2. Hexoses (glucose, fructose, galactose)
   C. Disaccharides
   D. Polysaccharides
      1. Starch and glycogen
      2. Cellulose
III. Lipids: Sometimes Too Much of a Good Thing
   A. Fats and Oils
      1. Definitions of lipids, fats, and oils
      2. Chemical structures
         a. Fatty acids
         b. Triglycerides
         c. Saturated, unsaturated, and polyunsaturated fats and oils
      3. Functions in living systems (insulation, cushioning, energy storage)
   B. Steroids
      1. Four-ring structure
      2. Cholesterol and atherosclerosis
IV. Proteins: Hair, Nails, Muscles, Skin, and Enzymes
   A. Nature and Formation of Proteins
   B. The Architecture of Proteins
      1. Amino acid monomer units
      2. Primary protein structure
      3. Three-dimensional structure
         a. Types of chemical bonds
         b. Importance in protein function
         c. Denaturation
V. Nucleic Acids: The Secret of Life?
   A. DNA and the Genetic Code
   B. Chemical structure
      1. Nucleotide monomer units
         a. Contain phosphate
         b. Contain ribose (in RNA) or deoxyribose (in DNA)

c. Contain purines (adenine, guanine) and pyrimidines (cytosine and either thymine--in DNA only--or uracil--in RNA only)

2. Double helix structure of DNA

C. DNA (genes) Specify Proteins Made in the Cell

Objectives

After you read and study the chapter [and the sections in brackets], you should be able to:

[C] 1. Identify the four major classes of molecules in living systems and the relationship in those substances between monomer units and biopolymers. [Section 9.1; Questions 1-3]

[C] 2. Distinguish among monosaccharides, disaccharides, and polysaccharides in terms of their chemical structures and functions in living systems. Cite specific examples of each type of carbohydrate. [Section 9.2; Questions 4-9]

[C] 3. Identify the following types of lipids: fatty acids, triglycerides, fats, oils, saturated lipids, unsaturated lipids, polyunsaturated lipids, and steroids. Cite their uses in the body. [Section 9.3; Questions 10-13]

[C] 4. Explain why proteins are among the most important chemicals in the body. Identify the structural features of amino acids and how they join together to make up the primary protein structure. [Section 9.4; Questions 14, 15, and 17]

[C] 5. Explain what bonds help maintain the three-dimensional structure of a protein, and how changes in primary structure affect the three-dimensional structure and function of proteins. [Section 9.4; Questions 16, 18, 19, and 23]

[C] 6. Distinguish between DNA and RNA in terms of their chemical structures and their functions in the cell. [Section 9.5; Questions 20-22]

Practice

<u>Objective 1:</u> Identify the four major classes of molecules in living systems and the relationship in those substances between monomer units and biopolymers.

Nucleic acids (DNA and RNA) are made from monomers called nucleotides.
Proteins are made from monomers called amino acids.
Carbohydrates include polysaccharides (starch, glycogen, and cellulose), which are made from monomers that are simple sugars, particularly glucose.
Lipids include triglycerides, which are made from simpler molecules called fatty acids and glycerol.
Figure 9.2 shows the four major types of biomolecules and their building block molecules.

Examples:

1. Insulin, a small protein hormone, consists of 51 amino acid units joined together. Most proteins contain more than 100 amino acid units.
2. Cellulose, a large carbohydrate molecule that provides most of the structural material in plants, is made from thousands of molecules of a simple sugar, glucose. Glucose itself is a carbohydrate, but (unlike cellulose) is not a macromolecule or biopolymer.
3. The smallest type of RNA, a nucleic acid, is made up of about 75-80 monomer units called nucleotides that join together. The largest type of nucleic acid, DNA, is made from as many as a million nucleotide units.
4. Fatty acids, glycerol, and steroids are lipids but are not biopolymers or macromolecules. Glycerol can join with three fatty acid molecules to form larger lipid molecules called triglycerides. Most of the fat in our bodies is triglyceride.

Drill:

1. Which of the macromolecules in Figure 9.2 is not a biopolymer?
2. Are all macromolecules biopolymers?
3. Are there any other types of chemicals besides those in Figure 9.2 that are necessary in living systems?

Objective 2: Distinguish among monosaccharides, disaccharides, and polysaccharides in terms of their chemical structures and functions in living systems. Cite specific examples of each type of carbohydrate.

Monosaccharides are carbohydrates containing one ring in their structure. They usually have a total of five or six carbon atoms, and specific examples are glucose, fructose, galactose, ribose, and deoxyribose.
Disaccharides are carbohydrates containing two ring structures joined together. They include sucrose, maltose, and lactose.
Polysaccharides are carbohydrates containing many (more than ten) ring structures joined together. They include glycogen, starch, and cellulose.

Examples:

1. Glucose is a monosaccharide. Two glucose units joined in a certain way make a disaccharide called maltose. Many glucose units joined in a certain way make the polysaccharide cellulose; joined in a different way, they make the polysaccharides glycogen and starch.
2. Fructose is a monosaccharide. One fructose molecule joined with one glucose molecule makes the disaccharide sucrose. Many fructose molecules joined together make a polysaccharide called insulin.

Drill:

1. Classify each of the following as a monosaccharide, disaccharide, or polysaccharide:
   a) celloboise (its formula is $C_{12}H_{22}O_{11}$)
   b) starch
   c) mannose (its formula is $C_6H_{12}O_6$)
2. Which of the following biomolecules always contains monosaccharide units: a) proteins, b) polysaccharides, c) disaccharides, d) lipids, e) nucleic acids
3. Name the most abundant monosaccharide in the human body.

Objective 3: Identify the following types of lipids: fatty acids, triglycerides, fats, oils, saturated lipids, unsaturated lipids, polyunsaturated lipids, and steroids. Cite their uses in the body.

Lipids are naturally oily or waxy materials that contain a large amount of hydrocarbon material and dissolve in nonpolar organic solvents better than in water.

Fatty acids contain a long (at least 12 carbon atoms) hydrocarbon chain and a carboxylic acid group ($-\overset{O}{\overset{\|}{C}}-O-H$) at the end.

Triglycerides consist of one molecule of glycerol joined to three fatty acid molecules by ester ($-\overset{O}{\overset{\|}{C}}-O-R$) bonds.

Fats are triglycerides that tend to be solids at room temperature.

Oils are triglycerides that tend to be liquids at room temperature.

Saturated lipids contain no carbon-to-carbon double (or triple) bonds.

Unsaturated lipids contain one carbon-to-carbon double (or triple) bond.

Polyunsaturated lipids contain more than one carbon-to-carbon double (or triple) bond.

Steroids have a complex, four-ring structure shown in Figure 9.11.

Examples:

1. $CH_3CH_2CH_2CH_2CH_2CH_2CH_2CH_2CH_2CH_2CH_2CH_2CH_2\overset{\displaystyle O}{\overset{\|}{C}}-O-H$ is a fatty acid. If the hydrocarbon chain were shorter (for example, 4 carbon atoms long) the molecule would be an organic (carboxylic) acid, but not a fatty acid.
2. The following molecule:

$$\begin{array}{l} CH_2-O-\overset{O}{\overset{\|}{C}}CH_2CH_2CH_2CH_2CH_2CH_2CH_2CH{=}CHCH_2CH{=}CHCH_2CH_2CH_2CH_2CH_3 \\ | \\ CH-O-\overset{O}{\overset{\|}{C}}CH_2CH_2CH_2CH_2CH_2CH_2CH_2CH_2CH_2CH_2CH_2CH_2CH_2CH_2CH_3 \\ | \\ CH_2-O-\overset{O}{\overset{\|}{C}}CH_2CH_2CH_2CH_2CH_2CH_2CH_2CH{=}CHCH_2CH{=}CHCH_2CH{=}CHCH_2CH_3 \end{array}$$

   is a triglyceride, an oil, and a polyunsaturated fat.
3. If the structure in Example 2 above were modified only to change each $-CH{=}CH-$ into $-CH_2CH_2-$, the molecule would be a triglyceride, a fat, and a saturated lipid.
4. The male sex hormone, testosterone, has the following structure:

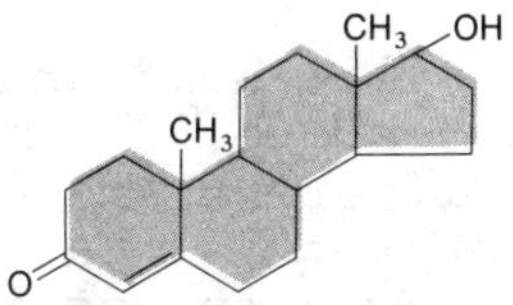

   This four-ring structure makes testosterone a steroid.

Drill:

1. In the structure in Example 2 above, the three carbon atoms at the left are in bold print. If they are numbered from top to bottom, which one--carbon number 1, 2, or 3--is joined by an ester linkage to a saturated fatty acid unit?
2. What are fats used for in the body?
3. Why is a fatty acid considered a lipid, but a short-chain organic acid is not a lipid?
4. What are steroids used for in the body?

Objective 4: Explain why proteins are among the most important chemicals in the body. Identify the structural features of amino acids and how they join together to make up the primary protein structure.

Proteins are structural components of hair, skin, nails, bone, connective tissue, and muscle. Proteins also (as enzymes) catalyze all the reactions in the body, (as antibodies) protect against infectious agents, and (as hormones) help regulate metabolism.

Amino acids are the monomer units in proteins. They contain a carboxylic acid (-C(=O)-O-H) group and an amino ($-NH_2$) group.

The carboxylic acid group of one amino acid joins with the amino group of another amino acid to form a peptide (amide) bond with the structure -C(=O)-N(H)-.

The sequence of amino acids joined together by peptide bonds in a protein is the protein's primary structure.

Examples:

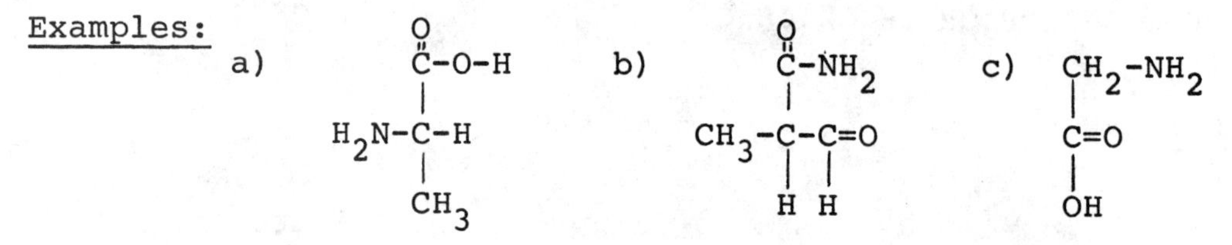

Substances a and c above are amino acids; they each have a carboxylic acid group and an amino group. Substance b is not an amino acid.

Drill:

1. Identify the carboxylic acid and amino functional groups in examples a and c above.
2. There are twenty different amino acids that can be in each position in a protein chain. For a sequence of two amino acid units, how many possible combinations of amino acids can occur? From this, do you think there are very many different possible primary protein structures (amino acid sequences) for a "typical" protein containing 100-200 amino acid units?
3. Name some important processes in the body that do not involve proteins in some way.

Objective 5: Explain what bonds help maintain the three-dimensional structure of a protein, and how changes in primary structure affect the three-dimensional structure and function of proteins.

Primary structure, the sequence of amino acids, is held together by covalent peptide (amide) bonds.

The three-dimensional structure (shape) of a protein depends on its primary structure and on other bonds that hold parts of the protein together. Those bonds include London forces, hydrogen bonds, disulfide bonds (covalent), and ionic bonds. Figure 9.14 illustrates this.

Denaturation is the disruption of the three-dimensional structure of a protein without disrupting its primary structure. But changes in primary structure also can change the shape of a protein. Since a protein needs a certain shape in order to function properly, denaturation or a change in primary structure often destroys protein function.

Examples:

1. Boiling an egg denatures proteins by disrupting hydrogen and ionic bonds. Without functioning proteins, eggs are no longer able to develop into chicks.
2. In sickle cell anemia, a change of one amino acid in the primary structure of a hemoglobin protein changes the shape and functioning ability of red blood cells.

Drill:

1. Reactions in the cell are catalyzed by proteins called enzymes. Why would those reactions cease at very high temperatures?
2. Strong acids can disrupt hydrogen bonds. Why do enzymes in saliva stop digesting food when the food reaches the stomach?
3. Mark the following true or false:
   a) Denaturation changes the primary structure of a protein.
   b) Sickle cell hemoglobin is an example of denatured protein.

Objective 6: Distinguish between DNA and RNA in terms of their chemical structures and their functions in the cell.

DNA (deoxyribonucleic acid) contains two purines (adenine, guanine), two pyrimidines (cytosine, thymine), deoxyribose, and phosphate. It carries the genetic information for what proteins the cell can make.

RNA (ribonucleic acid) contains two purines (adenine, guanine), two pyrimidines (cytosine, uracil), ribose, and phosphate. It helps convert the genetic information in DNA into proteins made by the cell.

Examples:

1. DNA contains thymine but not uracil; RNA contains uracil but not thymine.
2. DNA contains deoxyribose but not ribose; RNA contains ribose but not deoxyribose.
3. DNA normally exists as two strands wrapped around each other in a double helix; RNA normally is a single strand.
4. DNA carries the genetic information; RNA helps convert the genetic information into useful proteins.

Drill: Indicate whether the following are characteristics of DNA, RNA, or both:

1. Made of nucleotide monomer units
2. Typically occurs as a double helix
3. Contains a five-carbon sugar
4. Carries the genetic code
5. Does not normally contain thymine

Self-Test

---

1. The monomer units in collagen, the most abundant protein in our bodies, are a) monosaccharides, b) disaccharides, c) fatty acids, d) nucleotides, e) none of the above
2. The most abundant form of energy storage in the body is in the form of a) polysaccharide, b) triglyceride, c) steroid, d) protein, e) none of the above
3. Fill in the blanks appropriately:
   a) _______ sugar present in RNA
   b) _______ largest type of natural organic molecule
   c) _______ most abundant polysaccharide in plants
   d) _______ term for disruption of three-dimensional protein structure
   e) _______ type of fatty acid with no carbon-to-carbon double or triple bonds
   f) _______ triglyceride that is a liquid at room temperature
   g) _______ specific example of a disaccharide
   h) _______ type of bond joining two amino acids in primary protein structure
   i) _______ contains nucleotide monomer units
   j) _______ monomer unit in glycogen
4. Which of the following are not biopolymers: a) proteins, b) sucrose, c) triglycerides, d) DNA, e) cellulose
5. Steroids are lipids because they a) are made from fatty acids, b) are liquids at room temperature, c) have nonpolar, hydrocarbon characteristics, d) are a major form of energy storage, e) none of the above
6. Carbohydrates are important as a) structural materials for bone and skin, b) carriers of genetic information, c) sources of energy, d) hormones, e) none of the above
7. DNA differs from RNA because it a) is larger, b) contains guanine, c) is made from amino acids, d) contains ribose, e) carries genetic information for the cell
8. Sickle cell hemoglobin differs from normal hemoglobin in its a) incidence of peptide bonds, b) primary structure, c) number of amino acids, d) three-dimensional structure, e) none of the above

Answers

Objective 1:

1. lipids
2. No. Macromolecules such as nylon and various plastics (polyethylene, Teflon) are not biopolymers.
3. Yes. Living things also need water, vitamins, and minerals.

Objective 2:

1. a) disaccharide, b) polysaccharide, c) monosaccharide
2. b, c, and e (see Section 9.5)
3. glucose

Objective 3:

1. Carbon 2 is joined to a saturated fatty acid unit.
2. Insulation, cushioning, and energy storage.
3. A short-chain organic acid doesn't have enough hydrocarbon material to make it much more soluble in nonpolar organic solvents than in water (see the definition on p. 104 for a lipid). A fatty acid does have a long hydrocarbon chain.
4. Functions of steroids include male and female sex hormones, other hormones (such as for water and salt retention), and vitamin D (for maintenance of bone).

Objective 4:

1. The carboxylic acid group is on the top in a) and on the bottom in c), while the amino group is in the left center in a) and at the top right in c).
2. There are 400 possible two-amino-acid sequences. The number of amino acid sequences in a protein containing 100-200 amino acids is immense.
3. There aren't any.

Objective 5:

1. High temperatures disrupt the three-dimensional shapes of enzymes, so they no longer function as catalysts.
2. Stomach acid denatures the enzymes, so they can't function.
3. a) false, b) false

Objective 6:

1. both
2. DNA
3. both
4. DNA
5. RNA

Self-Test:

1) e; 2) b; 3) a) ribose, b) DNA, c) cellulose, d) denaturation, e) saturated, f) oil, g) sucrose, lactose, or maltose, h) peptide (amide), i) nucleic acids (DNA and RNA), j) glucose; 4) b and c; 5) c; 6) c; 7) a and e; 8) b and d

Evaluation:

If you missed more than one question on the self-test in any of the following groups, you need to review the section indicated:

Question Groups	Section
1, 3i, 4	9.1
3c, 3g, 3j, 6	9.2
2, 3e, 3f, 5	9.3
1, 3d, 3h, 8	9.4
3a, 3b, 3i, 7	9.5

Chapter 10

Soil and Mineral Resources
Using the Earth to Grow Food and Get Raw Materials

Outline

I. Chemistry of the Earth: Chemical Cycling and Energy Flow
   A. Our Life Support System
      1. Atmosphere
      2. Hydrosphere
      3. Lithosphere
      4. Biosphere
   B. Carbon and Oxygen Cycles
   C. Nitrogen Cycle
   D. Phosphorus Cycle

II. Resources: Transforming the Earth's Raw Materials into Useful Products
   A. What is a Resource?
   B. Types of Resources
      1. Nonrenewable resources
      2. Renewable resources
   C. Minerals and Ores
      1. Minerals
      2. Ores

III. Soil: A Vital and Abused Resource
   A. Soil: The Base of Life
   B. Soil Erosion

IV. Mineral and Metal Resources: Converting Sand to Computer Chips, and Getting Metals from Ores
   A. Silicon: An Abundant and Useful Metalloid Element
   B. Metallurgy: Converting Ores to Metals
   C. Removal of Additional Impurities Before or During Reduction
   D. Chemical Reduction of the Molten Ore
   E. Electrolytic Reduction of the Molten Ore

V. Environmental Consequences of Using Resources: Solid and Hazardous Wastes
   A. Environmental Impact of Resource Use
   B. Solid Wastes: Burn, Dump, Bury, Recycle, or Reuse?
   C. Hazardous Wastes: Not in *My* Backyard

Objectives

After you read and study the chapter [and the sections in brackets], you should be able to:

[C] 1. Describe how the carbon, oxygen, nitrogen, and phosphorus cycles affect and are affected by living things. [Section 10.1; Questions 1, 2, 3]

[C] 2. Give important characteristics of resources and distinguish between nonrenewable and renewable resources. [Section 10.2; Questions 4, 5, 6, 7]

[C] 3. Distinguish among the hard to reduce metals, the moderately hard to reduce metals, and the easy to reduce metals and describe the chemistry used to obtain the metals from their ores. [Section 10.4; Questions 8, 9, 10]

[M] 4. Describe the chemistry used to obtain pure silicon from its ores. [Section 10.4; Question 9]

[C] 5. Describe some environmental consequences of obtaining and using resources. [Sections 10.4, 5; Question 11]

[C] 6. Distinguish among the throwaway, recovery and recycling, and sustainable resources systems of waste management. [Section 10.5; Questions 12, 13, 15]

[M] 7. Describe advantages and disadvantages of the possible ways to manage hazardous wastes. [Section 10.5; Question 14]

Practice

Objective 1: Describe how the carbon, oxygen, nitrogen, and phosphorus cycles affect and are affected by living things.

The carbon and oxygen cycles are essentially inseparable. The following equations summarize the relationships: Photosynthesis (in green plants):

$$CO_2 + H_2O + \text{energy} \text{ ----> } \text{organic compounds} + O_2$$

Respiration and decay (in living plants and animals and in decay, and in combustion of fossil fuels):

$$\text{organic compounds} + O_2 \text{ ----> } CO_2 + H_2O + \text{energy}$$

The nitrogen cycle is more complex, but basically involves the conversion of N_2 from the air to nitrates in the soil by bacteria, algae, and lightning. Nitrates are taken up from the soil by plants, which form proteins, nucleic acids, and other compounds containing nitrogen which are essential to life. Animals get their nitrogen ultimately from plants. Some of this nitrogen eventually is converted to NH_3 and NH_4^+ in waste products and through decay. These are oxidized back to N_2. One possible sequence is:

$$N_2 \text{ ---> } NO_3^- \text{ ---> proteins ---> } NH_3 \text{ ---> } N_2$$

The phosphorus cycle also involves the uptake by plants of phosphates from the soil and water. Animals get their essential phosphorus ultimately from plants. Animal wastes and decaying plants and animals return this phosphorus to the soil and water.

<u>Drill:</u>

1. As you read this, you are a part of the carbon and oxygen cycles. How?
2. Describe how you participate in the nitrogen and phosphorus cycles.

<u>Objective 2:</u> Give important characteristics of resources and distinguish between nonrenewable and renewable resources.

Resources are forms of matter or energy that are:

(1) <u>Useful</u>. If something has no practical use, it is not considered a resource.

(2) <u>Inexpensive</u>. Our use of resources is controlled by their cost and their importance. Expensive resources will be used if no other alternative is available and if they are important to our way of life.

(3) <u>Plentiful</u>. This of course is related to expense. If we deplete our resources, they become too expensive to use.

(4) <u>Safe</u>. This is the ideal. However, our use of resources may result in environmental damage or health hazards, either known in advance or unforeseen.

Nonrenewable resources are those that are not being replaced (either by natural chemical cycles or by our own efforts), or are being replaced more slowly than they are being used.

Renewable resources have an unlimited supply or are replaced as they are used by natural chemical cycles or man's recycling efforts.

<u>Examples:</u>

1. Energy from flowing water. RENEWABLE--as long as rainfall isn't seriously disrupted by changing climate.
2. Gold. NONRENEWABLE--it is being mined from the ground but not replaced by natural cycles.

Drill:

1. There is more gold dissolved in the oceans than has ever been mined from the earth. Which of the four characteristics of a resource prevents this gold from being used?
2. Which of the four characteristics of a nonrenewable resource are likely to change as it is used over long periods of time?
3. Circle N for nonrenewable or R for renewable resources:

a.	solar energy	**N**	**R**
b.	fossil fuels	**N**	**R**
c.	wood	**N**	**R**
d.	fertile soil	**N**	**R**

Objective 3: Distinguish among the hard to reduce metals, the moderately hard to reduce metals, and the easy to reduce metals and describe the chemistry used to obtain the metals from their ores.

Easy to reduce metals either occur naturally in the metallic state (such as gold, silver, and platinum), so no reduction is required, or the ores can be reduced to the free metal simply by roasting (heating the ore in the presence of air). An example is

$$CuS + O_2 \longrightarrow Cu + SO_2$$

Notice that copper, silver, gold, platinum, and mercury are in the same area of the periodic table. This can help you remember them.

Moderately hard to reduce metals require a chemical reducing agent (see Section 6.3) to obtain the free metal. Impure carbon obtained from coal (called coke) or carbon monoxide is often used as the reducing agent. Typical equations are

$$Fe_2O_3 + 3\ C \longrightarrow 2\ Fe + 3\ CO$$

$$ZnO + C \longrightarrow Zn + CO$$

These metals (see Table 10.1) include tin and lead and several of the transition metals.

Hard to reduce metals require electrolytic reduction, which is done by passing an electric current through the molten ore. An example is:

$$2\ KCl\ (molten) \longrightarrow 2\ K + Cl_2$$

These metals include the alkali and alkaline earth metals, and aluminum (see Table 10.1).

Impurities often need to be removed from the ore before or during reduction, and from the free metal that results.

Many ores contain impurities that are removed by adding limestone ($CaCO_3$) during the reduction process. The limestone decomposes when heated, and then reacts with impurities such as SiO_2 as follows:

$$CaCO_3 \text{ ---> } CaO + CO_2$$

$$SiO_2 + CaO \text{ ---> } CaSiO_3 \text{ (\underline{slag})}$$

Carbon is a common impurity in the metals obtained by chemical reduction. It is removed by heating the metal in air or oxygen, which oxidizes the carbon to carbon monoxide and carbon dioxide.

Drill:

1. Find barium (Ba) on the periodic table. In which of the three categories of metals listed above would you place it?
2. Zinc is adjacent to copper on the periodic table, yet the chemistry involved in obtaining them from their ores is different. Starting with CuS and ZnS ores, use equations to summarize this chemistry (see page 260).
3. Write equations for the reduction of Cr_2O_3 and Al_2O_3 to the metals.

Objective 4: Describe the chemistry used to obtain pure silicon from its ores.

Reduction: Silicon ores (such as SiO_2) are reduced using carbon, just like the moderately hard to reduce metals.

$$SiO_2 + 2\ C \text{ ---> } Si + 2\ CO$$

Purification: The silicon is purified by converting it to a compound which is easily vaporized and then condensed, leaving the impurities behind (much like purifying water by distillation).

$$Si + 2\ Cl_2 \text{ ---> } SiCl_4$$

$$\text{impure } SiCl_4 \text{ liquid ---> pure } SiCl_4 \text{ gas}$$

$$\text{pure } SiCl_4 \text{ gas ---> pure } SiCl_4 \text{ liquid}$$

The $SiCl_4$ is reduced to pure Si with hydrogen:

$$SiCl_4 + 2\ H_2 \text{ ---> } Si + 4\ HCl$$

The silicon must be purified even more for use in the electronics industries. This is done by zone refining, in which a heater moves along a bar of silicon, melting it as it goes. After the heater has passed, the bar solidifies again, but the impurities move with the molten area. Successive passes of the heater result in very pure silicon on one end of the bar. See figure 10.13.

Drill:

1. Write the equations for the reactions involved in obtaining silicon from its ores. Practice until you can write them from memory.

Objective 5: Describe some environmental consequences of obtaining and using resources.

Mining can cause serious and permanent scarring of the landscape (see Figures 10.5 and 10.14) and can result in serious water pollution problems (see Chapter 13).

Reduction of ores often is accompanied by air pollution problems. SO_2 and CO are very harmful air pollutants (see Chapter 12) and are produced in roasting and other reduction processes.

Manufacturing and other uses of resources require energy (which is generated in ways that may damage the environment - see Chapter 11) and result in the production of wastes. Of course energy is also required in the reduction of ores, particularly the hard to reduce metals which require large amounts of electricity.

Use of a product means that more will need to be produced, and the cycle is repeated with further damage to the environment. Recycling of used products often eliminates much of the environmental cost of mining, reduction, and manufacturing.

Drill:

1. Match the following processes with its environmental consequences (there may be more than one):

a. Mining	1. Energy use
b. Reduction	2. Water pollution
c. Manufacturing	3. Air pollution
	4. Landscape scarring

Objective 6: Distinguish among the throwaway, recovery and recycling, and sustainable resources systems of waste management.

The throwaway system means just what it says. Used materials are simply thrown "away", which means they are dumped or buried somewhere. Combustible wastes are burned. Valuable resources are lost as wastes become scattered in landfills across the countryside.

The recovery and recycling system involves sorting wastes and returning them to the production cycle (hence the name recycle). Used metals and glass are melted and made into new products. This avoids the mining and reduction steps with their environmental problems, and extends the life of nonrenewable resources. Where recycling is not practical (perhaps with tires and plastics), the wastes can be burned to generate electricity or heat.

The sustainable resource system encourages reuse of resources instead of recycling or throwing them away. It would penalize products that were not reusable or had a limited lifetime, and would substitute reusable products for products that would have to be recycled or thrown away (such as using returnable glass bottles instead of plastic bottles).

Study Table 10.2 carefully to see the three systems compared for several waste items.

Drill:

1. Contrast the three systems of waste management with regard to a styrofoam plastic cup or fast food container.

Objective 7: Describe advantages and disadvantages of the possible ways to manage hazardous wastes.

Land disposal has been the cheapest and easiest way to dispose of wastes, and most wastes have been disposed of by this method. Disadvantages include the probability that leaks and contamination of surrounding areas and groundwater will result. Safe and secure disposal sites (see Figure 10.26) are very difficult to construct and are very expensive, so the advantages are lost when this type of disposal is done properly.

Conversion to less hazardous or nonhazardous forms has the advantage of greatly reducing the volume of hazardous wastes that need to be stored. Chemical reactions such as neutralization, oxidation, and precipitation can be used to treat the wastes. Disadvantages include the expense of building facilities to treat wastes in this way.

Incineration eliminates the need for disposal altogether. Although this sounds attractive, it is expensive and can cause serious air pollution when done improperly. When done on land, the heat could be used to generate electricity, but it would pose a danger for people living nearby. If done at sea, threats to human health are reduced and it is cheaper, but there is a danger to marine life.

Recycling and/or reuse of hazardous materials also eliminates the need for disposal and further reduces the amount of waste generated. However, it is expensive, and the extra handling of the wastes causes some concern for the environment.

Drill:

1. Prepare a table summarizing the advantages and disadvantages of the various hazardous waste disposal options.

Self-Test

---

1. True or false: soil is a renewable resource.
2. What system of solid waste is used for most product packaging?
   a) throwaway b) recycle c) sustainable d) other
3. Iron ore is _________ to reduce to iron metal/
   a) easy b) moderate c) difficult d) impossible
4. Which of the following is cycles is not affected by the activities of humans?
   a) carbon/oxygen b) nitrogen
   c) phosphorus d) none; humans affect them all
5. Photosynthesis is part of which cycle?
6. What would it take for radioactive nuclear waste to become a resource?
7. Which element is second only to oxygen in abundance in the earth's crust?
8. What is the environmental consequence of producing metals that are found principally as sulfides?
9. What constitutes a hazardous waste?
10. Is desert sand a resource?
11. What technique is used to refine silicon into the ultra-pure material needed in computer chips?

Answers

Objective 1:

1. The oxygen you breathe is oxidizing organic compounds, converting them to carbon dioxide and water. You use the energy to live and move around (and think).
2. Proteins and other components of the food you eat contain nitrogen and phosphorus. You return some of these elements to the environment through your waste products.

Objective 2:

1. (2) It is too expensive to extract gold from seawater.
2. They are likely to become less plentiful, which will increase their cost.

3. a) R; b) N; c) R; d) R

Objective 3:

1. Hard to reduce (like Ca, Mg, etc.).
2. $CuS + O_2 \text{ ---> } Cu + SO_2$ (roasting)

   $2\ ZnS + 3\ O_2 \text{ ---> } 2\ ZnO + 2\ SO_2$ (roasting)

   $ZnO + C \text{ ---> } Zn + CO$ (chemical reduction)

3. $2\ Al_2O_3 \text{ ---> } 4\ Al + 3\ O_2$ (electrolytic reduction)

   $Cr_2O_3 + 3\ C \text{ ---> } 2\ Cr + 3\ CO$ (chemical reduction)

Objective 5:

1. a) 2,4; b) 1,3; c) 1

Objective 6:

1. Throwaway: It would be dumped, buried, or burned.
   Recovery and recycling: It would be returned for recycling, or if burned, would be burned in an electricity-generating plant.
   Sustainable resource: It would not be used, or if used, would be taxed.

Self-Test:

1) true; 2) a; 3) b; 4) d; 5) carbon/oxygen; 6) a way to convert it into some useful product; 7) silicon; 8) air pollution from sulfur dioxide; 9) any discarded material that may pose a substantial threat to human health; 10) yes; 11) zone refining

Evaluation:

If you missed more than one question on the self-test in any of the following groups, you need to review the section indicated:

Question Groups	Section
4, 5	10.1
1, 6, 10	10.2
3, 7, 11	10.4
2, 8, 9	10.5

Chapter 11

Energy Resources: Keeping Things Cool, Warm, Lighted, and Moving

---

Outline

I. Types and Use of Energy Resources: Will There Be Enough?
   A. The Sun: Source of Energy for Life on Earth
   B. Primary Energy Resources
      1. Renewable
      2. Nonrenewable
   C. Energy Use
      1. Industrialized nations
      2. Less-developed nations
II. Some Energy Concepts: Energy Quality, Energy Efficiency, and Net Energy
   A. Energy Quality
      1. High-quality energy
      2. Low-quality energy
   B. Increasing Energy Efficiency
   C. Net Energy: It Takes Energy to Get Energy
      1. High net energy ratio sources
      2. Low net energy ratio sources
III. Fossil Fuels: How Long Will They Last?
   A. Are We Running Out?
   B. Heavy Oils from Oil Shale and Tar Sands
      1. Oil Shale
         a. Abundance
         b. Problems
      2. Tar Sands
         a. Abundance
         b. Problems
   C. Increased Use of Coal
         a. Abundance
         b. Problems
            (1) Dangerous to mine
            (2) Expensive to move
            (3) Dirty to burn
   D. Burning Coal More Cleanly and Efficiently
   E. Synfuels from Coal Gasification and Liquefaction
      1. Coal gasification
         a. Coal ----> $CO + H_2$
         b. Coal ----> synthetic natural gas
      2. Coal liquefaction
IV. Nuclear Energy: Bane or Blessing?
   A. Here to Stay?
   B. Nuclear Fission Reactors
      1. Heat exchange system
      2. Electricity generation
      3. Core

a. Fuel rods
b. Control rods
c. Moderator

C. The Nuclear Fuel Cycle
D. Problems with Nuclear Power
1. Power plant safety
2. Radioactive waste storage
3. Spread of nuclear weapons
4. Economics

E. Breeder Reactors
F. Nuclear Fusion

V. Energy from the Sun, Biomass, Wind, the Earth's Heat, and Hydrogen: Working with Nature
A. Direct Solar Energy: Some Advantages and Difficulties
1. Passive solar heating systems
2. Active solar heating systems
3. Solar furnaces
4. Solar cells

B. Indirect Solar Energy from Biomass
1. Direct use as fuel
2. Conversion to biofuels

C. Indirect Solar Energy from Wind
D. Geothermal Energy
E. Hydrogen Gas as a Fuel

VI. Our Energy Options: An Overview
A. The near future
B. The intermediate future
C. The distant future

Objectives

After you read and study the chapter [and the sections in brackets], you should be able to:

[C] 1. Distinguish renewable from nonrenewable energy sources and identify examples of each. [Sections 11.1, 11.4, 11.5]

[C] 2. Tell whether a particular energy source will be used more, less, or about the same in the future. [Section 11.1; Questions 3a, 3e, 7]

[C] 3. Give examples of high and low energy efficiency processes. [Section 11.2; Questions 3b, 3c, 4, 5]

[C] 4. Give examples of high and low net energy processes. [Section 11.2; Questions 1, 2, 12]

[M] 5. Describe how a nuclear fission reactor works. [Section 11.4; Question 9]

[M] 6. List the major advantages and disadvantages of important nonrenewable energy sources. [Sections 11.3, 11.4; Question 8]

[M] 7. List the major advantages and disadvantages of important renewable energy sources. [Section 11.5; Questions 6, 10, 11]

Practice

Objective 1: Distinguish renewable from nonrenewable energy sources and identify examples of each.

Renewable energy sources are permanent sources or can be replenished as they are used.
Nonrenewable energy sources are temporary sources and are depleted as they are used.

Examples:

1. Solar energy. RENEWABLE--a permanent source.
2. Petroleum. NONRENEWABLE--a temporary source, is depleted as it is used. The geological processes which resulted in the formation of petroleum deposits are no longer occurring on the earth.
3. Biomass. RENEWABLE--can be replenished with proper management.

Drill:

1. Circle **R** for renewable or **N** for nonrenewable.

a.	Coal.	**R**	**N**
b.	Wind.	**R**	**N**
c.	Tides.	**R**	**N**
d.	Nuclear fission.	**R**	**N**
e.	Water power.	**R**	**N**
f.	Oil shale.	**R**	**N**

2. Can an energy source normally classified as renewable become nonrenewable?

Objective 2: Tell whether a particular energy source will be used more, less, or about the same in the future.

Study Figure 11.4. Notice how much each energy source contributed to our overall energy consumption in 1984. Compare with the projected values for the year 2000.

Drill:

1. Circle **I** if an increase in use is predicted, circle D if a decrease is predicted, and circle **U** if the use is predicted to be unchanged.

a)	Oil	**I**	**D**	**U**
b)	Natural gas	**I**	**D**	**U**
c)	Coal	**I**	**D**	**U**
d)	Nuclear	**I**	**D**	**U**
e)	Hydropower	**I**	**D**	**U**
f)	Biomass	**I**	**D**	**U**
g)	Solar, wind, geothermal	**I**	**D**	**U**

Objective 3: Give examples of high and low energy efficiency processes.

Energy efficiency is the amount of useful work done as a percentage of the total energy input. According to the second law of thermodynamics, no energy conversion process can be 100 % efficient because of the entropy increase that results.

Examples:

Study Table 11.1, which has examples of energy efficiencies from 14 % to 98 %.

Drill:

1. What is the most efficient way of generating electricity?
   a) nuclear power plant
   b) coal power plant
   c) hydroelectric power plant
2. What is the most efficient way to heat your home?
   a) a wood stove
   b) passive solar heat
   c) active solar heat
   d) an oil furnace
   e) natural gas
3. Which of the following is the most efficient way of heating your home with electricity?
   a) resistance heating
   b) a heat pump

Objective 4: Give examples of high and low net energy processes.

Net energy is the total energy obtained from a resource minus the energy needed to find, process, and transport it. According to the first law of thermodynamics, or law of conservation of energy, this equation must be true: total energy = net energy + energy used in obtaining the total energy. It is often expressed as a net energy ratio, which is the total energy divided by the energy used in producing it.

Example:

A nonrenewable energy source produces 10 units of energy for every 7 units of energy spent in mining it, processing it, and transporting it. The net energy is 3 units, and the net energy ratio is $10/7 = 1.4$. If this resource becomes depleted and more difficult to obtain, the energy spent in producing it might rise to 8 units, and the net energy ratio would drop to $10/8 = 1.25$.

Drill:

1. Modern agricultural and processing practices spend 9 calories of energy to produce 10 calories worth of food. What is:
   a) the net energy in this example?
   b) the net energy ratio of modern agriculture?

2. Primitive agricultural practices, with little processing or transportation costs, can produce up to 50 calories of food energy by spending only 1 calorie worth of energy. What is:
   a) the net energy in this example?
   b) the net energy ratio of primitive agriculture?
3. Which of these has the highest net energy ratio in transportation?
   a) alcohol b) gasoline
   c) natural gas d) coal liquefaction
4. Which of these has the highest net energy ratio in space heating?
   a) oil b) electric heat
   c) active solar d) passive solar

Objective 5: Describe how a nuclear fission reactor works.

In the core of a nuclear reactor, heat energy is produced by a nuclear fission reaction. The fuel for the reaction is contained within long, thin fuel rods. Neutrons produced by a fission reaction in one rod travel into adjacent rods where they initiate other fission reactions. In between fuel rods is found the moderator (usually water), which slows down neutrons and makes them more efficient in continuing the fission process. Control rods are also present within the core. These rods absorb neutrons, preventing them from continuing the fission process. Partially removing the control rods from the core increases the rate of the fission reaction, which generates more heat. With the control rods completely within the core, the fission reaction stops and the reactor is shut down.

The water also functions to carry heat away from the core. In a heat exchanger the heat from the water is transferred to water in another loop, causing it to boil. The steam is used to spin the blades of a turbine which is connected to the generator.

Drill:

1. The fission reaction takes place within the:
   a) control rods b) moderator
   c) fuel rods d) heat exchanger
2. The product of the fission reaction which is used to generate electricity is:
   a) neutrons b) radiation
   c) water d) heat energy
3. Neutrons are slowed down and made more efficient by:
   a) the control rods b) the moderator
   c) the fuel rods d) the heat exchanger
4. Neutrons are absorbed by the:
   a) moderator b) fuel rods
   c) control rods d) heat exchanger

Objective 6: List the major advantages and disadvantages of important nonrenewable energy sources.

Drill:
Cut out the flash cards and memorize the material on them.

Objective 7: List the major advantages and disadvantages of important renewable energy sources.

Drill:
Cut out the flash cards and memorize the material on them.

Self-Test

1. What is one advantage of nuclear fusion as an energy source?
2. Is hydropower renewable or nonrenewable?
3. Rank the following three methods of home heating from highest net energy efficiency to lowest:
   a) wood stove b) electric heat c) gas furnace
4. Which of the following fossil fuels is in greatest abundance?
   a) oil b) gas c) coal d) all equal
5. What is biomass?
6. Which of the following energy systems currently has the lowest net energy ratio?
   a) nuclear b) solar c) fossil fuel d) all equal
7. What is a synfuel?
8. In a fission reactor, nuclear energy is first converted into
   a) light b) heat c) radiation d) motion
9. Is use of nuclear power expected to increase or decrease in the future?
10. Is the use of photovoltaic cells to produce electricity an example of passive or active solar energy production?
11. Choose the renewable energy resource:
    a) coal b) geothermal c) oil d) gas
12. Wind energy can be produced without any adverse environmental impact. True or false?

Answers

Objective 1:

1. a) N; b) R; c) R; d) N; e) R; f) N
2. Yes, with mismanagement. Trees, if overharvested, can become depleted.

Objective 2:

1. a) D; b) D; c) I; d) I; e) U; f) I; g) I

Objective 3:

1. c
2. b
3. b

Objective 4:

1. a) 1 calorie; b) 1.1
2. a) 49 calories; b) 50
3. c
4. d

Flash Cards for Objectives 6 and 7

1. Coal — Nonrenewable	2. Solar — Renewable
3. Petroleum and natural gas — Nonrenewable	4. Biomass — Renewable
5. Oil shale and tar sands — Nonrenewable	6. Wind — Renewable
7. Nuclear fission — Nonrenewable	8. Geothermal — Renewable
9. Nuclear fusion — Nonrenewable	10. Water power — Renewable

2. Advantages:
   - Low environmental cost
   - Abundant
   - Moderate to high net energy

   Disadvantages:
   - Sporadically available
   - Moderate to high cost

4. Advantages:
   - Renewable with proper management
   - A use for organic waste

   Disadvantages:
   - Moderate environmental cost
   - Low net energy

6. Advantages:
   - Low environmental cost
   - Moderate net energy

   Disadvantages:
   - Low large-scale availability
   - Sporadically available

8. Advantages:
   - Alternative to nonrenewable sources

   Disadvantages:
   - Low availability
   - Low net energy
   - Moderate environmental cost

10. Advantages:
    - High net energy
    - Low environmental cost

    Disadvantages:
    - Low availability

1. Advantages:
   - Abundant
   - High net energy

   Disadvantages:
   - Hazardous to mine
   - High transport cost
   - High environmental cost

3. Advantages:
   - Readily available now
   - High net energy

   Disadvantages:
   - Low future availability
   - Dependence on imports

5. Advantages:
   - Large deposits

   Disadvantages:
   - High cost
   - Low net energy
   - High environmental cost

7. Advantages:
   - Lowers dependence on fossil fuels

   Disadvantages:
   - High cost
   - Radioactive wastes
   - Low net energy

9. Advantages:
   - Virtually unlimited fuel
   - No radioactive waste

   Disadvantages:
   - Not now available

Objective 5:

1. c
2. d
3. b
4. c

Self-Test:

1) unlimited fuel or lack of radioactive waste; 2) renewable; 3) c > a > b; 4) c; 5) organic matter such as wood or crop wastes that can be burned directly; 6) a; 7) synthetic gas or oil; 8) b; 9) increase; 10) active; 11) b; 12) false

Evaluation:

If you missed more than one question on the self-test in any of the following groups, you need to review the section indicated:

Question Groups	Section
2, 9, 11	11.1
3, 6	11.2
4, 7	11.3
1, 8	11.4
5, 10, 12	11.5

Chapter 12

Air Resources and Air Pollution: Keeping Breathing Safe

Outline

I. Major Air Pollutants: Types and Sources
  A. Our Air Resources: The Atmosphere
    1. The troposphere
    2. The stratosphere
  B. Our Polluted Air
  C. Major Pollutants and Sources
    1. Primary air pollutants
    2. Secondary air pollutants
  D. Types of Smog
    1. Industrial smog
    2. Photochemical smog
  E. Indoor Air Pollution
  F. Heat: The Ultimate Pollutant
  G. Local Climate, Topography, and Air Pollution Intensity

II. Industrial Smog and Acid Deposition: Are Fish and Trees Safe?
  A. The Formation of Industrial Smog
  B. Acid Deposition
  C. Controlling Industrial Smog and Acid Deposition
    1. Sulfur oxides
    2. Particulate material

III. Photochemical Smog: Cars + Sunlight = Tears
  A. The Chemistry of Photochemical Smog
    1. Nitrogen oxides
    2. Ozone
    3. PANs and other organics
  B. Controlling Pollution from Automobiles
    1. Air to fuel ratios
    2. Afterburners
    3. Catalytic converters
    4. Modified engines
  C. Lead Pollution

IV. Some Effects of Air Pollution: Damaging Materials, Plants, and People and Altering the Ozone Layer and Global Climate
  A. Damage to Property and Plants
    1. Erosion, corrosion, and soiling of surfaces
    2. Stunting and destruction of plants
  B. Damage to Human Health
  C. Are We Depleting the Ozone Layer?
  D. Are We Changing the Earth's Climate?
    1. The greenhouse effect
    2. The effect of particulate material

V. Summary of Air Pollution Problems and Their Control: Difficult But Not Impossible

Objectives

[C] 1. Distinguish primary air pollutants from secondary air pollutants and give examples of each. [Section 12.1; Question 3]

[C] 2. Tell which of the important primary and secondary pollutants come from motor vehicle exhaust and which come from oil or coal burning. [Sections 12.1, 2, 3; Questions 2, 5a, 5b, 11]

[C] 3. Distinguish industrial smog from photochemical smog. [Sections 12.1, 2, 3; Question 2]

[C] 4. Know the relationships between air pollution and climate. [Sections 12.1, 2, 3, 4; Questions 1, 2, 7]

[C] 5. Describe the major health effects of industrial smog and photochemical smog. [Sections 12.1, 2, 3, 4; Question 2]

[C] 6. Describe methods of controlling industrial and photochemical smog. [Sections 12.2, 3, 5; Questions 2, 4, 5c, 6, 8, 10]

Practice

Objective 1: Distinguish primary air pollutants from secondary air pollutants and give examples of each.

Primary air pollutants are chemicals that enter the atmosphere directly as a result of some process or activity.

Secondary air pollutants are formed in the atmosphere through chemical reactions among primary pollutants, or between primary pollutants and normal components of the atmosphere.

Examples:

1. Carbon monoxide (CO). PRIMARY--enters the atmosphere by way of exhaust gases from automobile engines.
2. Sulfur dioxide (SO_2). PRIMARY--enters the atmosphere wherever coal is burned.
3. Sulfur trioxide (SO_3). SECONDARY--formed in the atmosphere by reaction between SO_2 and oxygen.
4. Ozone (O_3). SECONDARY--formed in the atmosphere when nitrogen dioxide (NO_2) is exposed to sunlight.
5. Nitric oxide (NO). PRIMARY--formed in automobile engines and other places where air (a mixture of nitrogen and oxygen) is subjected to very high temperatures.
6. Sulfuric acid (H_2SO_4). SECONDARY--formed when sulfur trioxide (SO_3) reacts with water in the atmosphere.

Drill:

1. Circle **P** for primary or **S** for secondary pollutants.

a.	Hydrocarbons.	**P**	**S**
b.	Peroxyacyl nitrates (PANs)	**P**	**S**
c.	Soot (particulate matter).	**P**	**S**
d.	Nitric acid (HNO_3).	**P**	**S**
e.	Nitrogen dioxide (NO_2).	**P**	**S**
f.	Formaldeyde (CH_2O).	**P**	**S**
g.	Lead compounds ($PbBr_2$, etc.).	**P**	**S**

Objective 2: Tell which of the important primary and secondary pollutants come from motor vehicle exhaust and which come from oil or coal burning.

Primary pollutants: Automobiles and other vehicles are major sources of nitric oxide (NO), carbon monoxide (CO), hydrocarbons, and lead compounds. Burning coal and oil in industry, in electric power production, and in home heating is the major source of sulfur dioxide (SO_2) and particulate matter (such as soot).

Secondary pollutants: The following reactions occur after primary pollutants from automobiles enter the atmosphere:

$2\ NO + O_2 \longrightarrow 2\ NO_2$ (brown)

$NO_2 + H_2O \longrightarrow NO + HNO_3$ (nitric acid)

$HNO_3 + NH_3 \longrightarrow NH_4NO_3$ (ammonium nitrate)

$NO_2 \longrightarrow NO + O$ (on exposure to sunlight)

$O_2 + O \longrightarrow O_3$

In addition, ozone (O_3), nitrogen oxides, and hydrocarbons react in the presence of sunlight to form complex materials like peroxyacyl nitrates (PANs), formaldehyde (CH_2O), and other products. The nitric acid and ammonium nitrate are important components of acid deposition, more commonly known as acid rain.

The following reactions occur after primary pollutants from coal and oil burning enter the atmosphere:

$2\ SO_2 + O_2 \longrightarrow 2\ SO_3$

$SO_3 + H_2O \longrightarrow H_2SO_4$ (sulfuric acid)

$H_2SO_4 + 2\ NH_3 \longrightarrow (NH_4)_2SO_4$ (ammonium sulfate)

Sulfuric acid and ammonium sulfate are the most important sources of acid deposition (acid rain).

Drill:

1. Circle **M** for motor vehicles or **C** for coal or oil burning as the major source of the following pollutants.

a.	Carbon monoxide (CO).	**M**	**C**
b.	Nitric acid (HNO_3).	**M**	**C**
c.	Sulfur dioxide (SO_2).	**M**	**C**
d.	Ozone (O_3).	**M**	**C**
e.	Sulfuric acid (H_2SO_4).	**M**	**C**
f.	Ammonium nitrate (NH_4NO_3).	**M**	**C**
g.	Lead compounds ($PbBr_2$, etc.)	**M**	**C**
h.	Soot.	**M**	**C**

2. Pollutants containing sulfur come primarily from:
   a. motor vehicles. b. coal or oil burning.
3. Most of the pollutants containing nitrogen come primarily from:
   a. motor vehicles. b. coal or oil burning.
4. Most of the acid deposition (acid rain) comes from:
   a. motor vehicles. b. coal or oil burning.

Objective 3: Distinguish industrial smog from photochemical smog.

Industrial smog, also called grey smog, has high levels of sulfur oxides and particulate matter from coal and oil burning. It is typically found in cities with heavy industry (iron and steel making, petroleum refining, etc.), and where coal and oil is burned in large quantities for electric power and heating. Cool, damp weather increases the severity of this type of smog because of increased fuel consumption for heating and because of the increased rate of formation of sulfuric acid in the damp air.

Photochemical smog, also called brown smog, has high levels of carbon monoxide, nitrogen oxides, ozone, and other secondary pollutants from automobile exhaust. Sunlight increases the rate of formation of many of the secondary pollutants, so sunny weather brings the most severe episodes of this type of smog.

Drill:

1. Circle I for industrial smog or **P** for photochemical smog as the predominant type of smog in:

a.	Los Angeles	I	**P**
b.	Pittsburgh	I	**P**
c.	Chicago	I	**P**
d.	Denver	I	**P**
e.	Salt Lake City	I	**P**
f.	Philadelphia	I	**P**

2. Photochemical smog would predominate in:
   a. winter. b. summer.
3. a. What gives the brown color to photochemical smog and the gray color to industrial smog?

b. What gives the gray color to industrial smog?

4. A sample of rainwater is analyzed and found to have relatively high concentrations of sulfuric acid and ammonium sulfate. Which type of smog would have produced this acid rain?

Objective 4: Know the relationships between air pollution and climate.

Air pollution as a result of climate. As was mentioned under the previous objective, climate has an effect on the type and severity of air pollution. Industrial smog is worst in cool, damp weather. Industrialized cities with wet winter climates have their worst episodes of smog in the winter months, particularly in the early morning hours. Photochemical smog is worst in sunny weather since the reactions that form the secondary pollutants occur in the presence of sunlight. This type of smog is worst during the summer months, particularly around noon when the sun is high.

Both types of smog are made much worse during thermal inversions. Inversions occur when cold, dense air is trapped by a layer of warmer air. Pollutants accumulate in the cold air, which does not rise and disperse the pollutants. Study Figure 12.4, which compares the normal pattern with the inversion pattern.

Climate changes as a result of air pollution. The greenhouse effect is the increased warming of the surface of the earth caused by carbon dioxide and water vapor in the atmosphere. Carbon dioxide is produced whenever coal, oil, gasoline, wood, or any other organic material is burned. Not normally considered a pollutant, the CO_2 level in the atmosphere is increasing because of man's activities. This may result in serious changes in the earth's climate caused by increased absorption of the sun's radiation.

Perhaps counteracting the effect of CO_2 is the smoke, soot, and other particulate matter we inject into the atmosphere. This material reflects some of the sun's radiation, preventing it from reaching the earth. Scientists do not agree on the overall effect on the earth's future climate that these pollutants will have.

Even though it does not affect climate, the ozone layer in the stratosphere affects us greatly by absorbing most of the sun's ultraviolet radiation. Many of our pollutants may reach the stratosphere where they will react with the ozone, decreasing its shielding effect. Increased ultraviolet radiation would mean more sunburn, skin cancer, crop damage, and perhaps even more serious effects than these.

Drill:

1. Which type of smog is made worse by thermal inversions?
2. a) A thermal inversion consists of a layer of _______ air next to the ground with a layer of _______ air above it. (Fill in the blanks with "warm" or "cold").
   b) The normal pattern consists of a layer of _______ air next to the ground with _______ air above it.
3. Circle **I** for industrial smog or **P** for photochemical smog.
   a. Worst during winter months. **I** **P**
   b. Worst around noon. **I** **P**
   c. Worst in the early morning. **I** **P**
   d. Worst during summer months. **I** **P**
4. What components of the atmosphere cause the greenhouse effect?
   a. carbon dioxide b. ozone c. water vapor
5. Considering the greenhouse effect, increased levels of carbon dioxide in the atmosphere should result in:
   a. a warmer climate. b. a cooler climate.
6. Increased levels of particulate matter should result in:
   a. a warmer climate. b. a cooler climate.
7. Why is ozone both a beneficial and a harmful component of the atmosphere?

Objective 5: Describe the major health effects of industrial smog and photochemical smog.

Industrial smog: Sulfur oxides and sulfuric acid irritate air passages and lungs and can lead to chronic bronchitis and aggravation of colds, pneumonia, and asthma. In addition to irritating lungs, the particulate matter in industrial smog can contribute to the development of lung cancer and emphysema.

Photochemical smog: High carbon monoxide levels from automobile exhaust cause headaches, fatigue, and impaired judgment (very high amounts leading to rapid death). Lead compounds in automobile exhaust can lead to high blood pressure, heart attacks, and strokes. Ozone and other secondary pollutants in photochemical smog cause burning, watery eyes.

Drill:

1. Circle **I** for industrial smog or **P** for photochemical smog as a possible cause of the following health problems.
   a. Difficulty breathing. **I** **P**
   b. Mental confusion while driving. **I** **P**
   c. High blood pressure. **I** **P**
   d. Burning eyes. **I** **P**
   e. Sore throat. **I** **P**
2. Name pollutants that might cause each of the health problems in question 1.

Objective 6: Describe methods of controlling industrial and photochemical smog.

Industrial smog: The impact of sulfur oxides and particulate matter can be reduced in three basic ways.
1) Produce less of them in the first place. Energy conservation and alternate energy sources would allow us to burn less fossil fuels. Using low-sulfur fuels or removing the sulfur before combustion would reduce the amount of sulfur oxides produced. Fluidized-bed combustion is a more efficient way of burning coal, producing less particulate matter. Converting coal to gaseous or liquid fuels (synfuels) that contain much less sulfur is also a possibility.
2) Prevent them from leaving the smokestack. Fluidized-bed combustion traps sulfur oxides by converting them to calcium sulfate ($CaSO_4$) or calcium sulfite ($CaSO_3$). Scrubbers use water and sometimes limestone to trap both sulfur oxides and particulate matter. Electrostatic precipitators, baghouse filters, and cyclone separators all remove particulate matter by electrical attraction, filtering, or swirling, respectively.
3) Counteract the damage done after they leave the smokestack. This is the most difficult. Some attempts have been made to neutralize the acidity in lakes and soils affected by acid deposition.

Photochemical smog: The impact of automobile exhaust and related pollutants can be reduced in the following ways:
1) Use other fuels. Alcohol fuels, propane, and hydrogen are fuels which burn more cleanly than gasoline and will find greater use in the future. Unleaded gasoline will reduce lead pollution, but the price of unleaded gasoline is often higher than leaded gasoline, and it contains greater amounts of known or suspected carcinogens.
2) Modify or eliminate the internal combustion engine. Cars that use solar power, electric batteries, and fuel cells will become available in the future. Other engine designs, such as the stratified-charge engine, the Rankine engine, and the Stirling engine, are more efficient, resulting in better gas mileage and lower emissions.
3) With existing engines and fuels, find ways to reduce exhaust emissions. Catalytic converters cause carbon monoxide and hydrocarbons to react with oxygen, lowering the amounts emitted into the air. Afterburners recirculate exhaust gases, causing more complete combustion but increasing nitrogen oxide emissions.
4) Adjust the fuel-to-air ratio. A lean mixture (more air, less fuel) lowers emissions of carbon monoxide and hydrocarbons, but increases emission of

nitrogen oxides. A very lean mixture reduces nitrogen oxide emissions, but the engine misfires. A rich mixture (less air, more fuel) reduces nitrogen oxides, but increases emission of carbon monoxide and hydrocarbons. A properly adjusted carburetor is necessary to achieve the best balance.

Drill:

1. A car adjusted to high altitude driving will misfire when driven at low altitudes. Why?
2. Afterburners operate at much higher temperatures than catalytic converters. What pollutant is increased as a result?
3. Why can't gasoline containing lead be used in cars equipped with catalytic converters?
4. What advantage does a scrubber have over an electrostatic precipitator for cleaning coal combustion gases?

Self-Test

---

1. Automobile exhaust is a ________ (primary, secondary) air pollutant.
2. Which of the following is the major air pollutant emitted from coal-fired power plants?
   a) CO b) PAN c) NO d) SO_2
3. Which one of the following is a secondary air pollutant?
   a) CO b) PAN c) NO d) SO_2
4. Is there more danger from air pollution when a layer of warm air or cool air lies above a city's atmosphere?
5. Photochemical smog is a primary air pollutant. True or false?
6. How do peroxyacyl nitrates (PANs) get into the air?
   a) industrial emmissions b) acid rain
   c) automobile exhaust d) volcanic eruptions
7. Does acid rain come more from the air pollution caused by industry or motor vehicles?
8. What air pollutant damages marble statues?
   a) CO b) PAN c) NO d) SO_2
9. Ozone is a ________ (protectant, pollutant) in the stratosphere, but it is a __________ in the troposphere.
10. True or false, carbon dioxide, though non-toxic, may pollute the atmosphere by changing the climate.
11. Which of the following combines most readily with the hemoglobin in your blood?
    a) CO b) PAN c) NO d) SO_2
12. What metal in automobile exhaust causes air pollution?
13. Scrubbers, using a slurry of limestone and water, cleanse smokestack emissions of which pollutant?
    a) CO b) PAN c) NO d) SO_2

Answers

Objective 1:

1. a) P; b) S; c) P; d) S; e) S; f) S; g) P

Objective 2:

1. a) M; b) M; c) C; d) M; e) C; f) M; g) M; h) C
2. b
3. a
4. b

Objective 3:

1. a) P; b) I; c) I; d) P; e) P; f) I
2. b
3. a) Nitrogen dioxide.
   b) Soot (along with damp, foggy conditions).
4. Industrial smog. Photochemical smog would have produced more nitric acid and ammonium nitrate.

Objective 4:

1. Both industrial and photochemical smog
2. a) cold, warm; b) warm, cold
3. a) I; b) P; c) I; d) P
4. a and c
5. a
6. b
7. It is beneficial in the stratosphere because it screens out ultraviolet radiation. At lower levels it is a component of photochemical smog, irritating eyes and causing damage to crops and other materials.

Objective 5:

1. a) I; b) P; c) P; d) P; e) I
2. a) Sulfur dioxide, sulfuric acid.
   b) Carbon monoxide.
   c) Lead compounds.
   d) Ozone, PANs.
   e) Sulfur dioxide, sulfuric acid.

Objective 6:

1. More oxygen is present in a given volume of air at lower altitudes, causing misfire.
2. Nitrogen oxides.
3. Lead ruins the catalyst materials.
4. It removes sulfur oxides as well as particulate matter.

Self Test:

1) primary; 2) d; 3) b; 4) warm; 5) false; 6) c; 7) industry; 8) d; 9) protectant, pollutant; 10) true; 11) a; 12) lead; 13) d

Evaluation:

If you missed more than one question on the self-test in any of the following groups, you need to review the section indicated:

Question Groups	Section
1, 3, 4, 5	12.1
2, 7, 13	12.2
6, 9, 12	12.3
8, 10, 11	12.4

Chapter 13

Water Resources and Water Pollution: Keeping Drinking Safe

Outline

I. Water: A Unique Molecule
   A. Properties of Water
      1. Heat capacity
      2. Heat of vaporization
      3. Density of solid and liquid
      4. Freezing and boiling points
      5. Surface tension
      6. Wetting ability
   B. Molecular Structure and Hydrogen Bonding
II. The Earth's Water Resources: Will There Be Enough?
   A. World Water Resources
   B. The Hydrologic Cycle
      1. Surface water
      2. Groundwater
   C. Do We Have a Water Shortage?
III. Major Water Pollutants: Types, Sources, and Effects
   A. What is Water Pollution?
   B. Types and Sources of Water Pollutants
      1. Oxygen-demanding wastes
      2. Disease-causing agents
      3. Inorganic chemicals
      4. Synthetic organic chemicals
      5. Plant nutrients
      6. Sediments
      7. Radioactive substances
      8. Heat
   C. Effects of Water Pollutants
      1. Nuisance and aesthetic insult
      2. Property damage
      3. Damage to plant and animal life
      4. Damage to human health
      5. Major ecosystem disruption
IV. Organic Pollutants and Dissolved Oxygen: Keeping Fish Alive
   A. Dissolved Oxygen and Biochemical Oxygen Demand
      1. Aerobic decay
      2. Anaerobic decay
      3. Biochemical oxygen demand
   B. Rivers and Lakes: A Comparison
      1. Rivers
      2. Lakes and eutrophication
V. Inorganic Pollutants: Acids, Salts, and Toxic Metals
   A. Acidity
      1. Acid deposition
      2. Industrial activities

        3. Abandoned mines
    B. Salinity
        1. Sources of salinity
        2. Desalination
    C. Toxic Metals: Cadmium and Mercury
VI. Groundwater Pollution: Contaminating Our Drinking Water
    A. A Growing Problem
    B. Control of Groundwater Pollution
VII. Water Pollution Control: Chemical and Ecological Approaches
    A. Sewage Treatment
        1. Primary sewage treatment
        2. Secondary sewage treatment
        3. Tertiary sewage treatment
            a. Precipitation
            b. Filtration
            c. Reverse osmosis
        4. Chlorination
    B. Ecological Waste Management and Recycling
    C. Recent Water Pollution Laws and Control Efforts in the United States

---

Objectives

After you read and study the chapter [and the sections in brackets], you should be able to:

[M] 1. List important physical properties of water and explain them in terms of its molecular structure and hydrogen bonding. [Section 13.1; Questions 1, 2, 3, 4]

[M] 2. List the major sources and effects of important water pollutants. [Sections 13.3, 13.5; Questions 5b, c, d, 9, 10]

[M] 3. Explain the differing solubilities of oxygen and salts in water, and how solubility changes as a function of temperature. [Sections 13.3, 13.5; Questions 8, 13, 14]

[C] 4. Distinguish between dissolved oxygen (DO) and biochemical oxygen demand (BOD), and know how they relate to water quality. [Section 13.4; Question 11]

[C] 5. Distinguish between aerobic and anaerobic decay. [Section 13.4; Question 7]

[C] 6. Explain how rivers and lakes differ in their pollution problems. [Section 13.4; Questions 5a, e]

[C] 7. Distinguish among primary, secondary, and tertiary sewage treatments. [Section 13.7; Question 12]

Practice

Objective 1: List important physical properties of water and explain them in terms of its molecular structure and hydrogen bonding.

Hydrogen bonding: Review the material in Section 4.6 on hydrogen bonds. Recall that a hydrogen bond is an attraction between a hydrogen atom (that is bonded to F, O, or N) and an atom of F, O, or N in another molecule. In water it is the attraction of a hydrogen atom in one water molecule to an oxygen atom in another water molecule. It is a strong attraction.

Molecular structure: Figures 4.17 and 13.1 illustrate the bent or angular shape of water molecules. This shape and the very polar H-O bond cause the attractive forces between water molecules to be unusually strong. If the molecule were linear instead of angular, it would be nonpolar and the attractive forces would be much weaker.

Physical properties: Water has unusually high melting and boiling points compared to compounds containing molecules of similar size and weight. Its heat of vaporization (the heat needed to vaporize a given mass of it) and its heat capacity (the heat needed to change the temperature of a given mass of it) also are unusually high. These properties all are accounted for by the strong attractive forces between water molecules. Much energy is required to permit the molecules to break free from each other.

Water's high surface tension results from these strong attractive forces also. Water molecules at the surface are pulled inward strongly. If water is placed on a surface to which it is not attracted, it will form droplets or "beads" because of this inward attraction.

Because of their shape, water molecules crystallize in a hexagonal pattern as shown in Figure 13.2. This causes the unique shapes of snowflakes. It is responsible also for the expansion of water as it freezes (notice the large hole in the middle of the hexagon), resulting in ice with a lower density than that of water.

Drill:

1. H_2S is a gas at room temperature, whereas H_2O is a liquid. Why?
2. CO_2 is a gas at room temperature, whereas H_2O is a liquid. Why?
3. Mars has polar "ice caps" consisting of solid CO_2 (dry ice). If it "snows" on Mars, do you think the "snow-flakes" are six-sided? Why?

4. The molecules in gasoline are nonpolar and do not form hydrogen bonds. Circle **W** for water of G for gasoline.

a.	Has a higher melting point.	**W**	G
b.	Has a higher boiling point.	**W**	G
c.	Has a lower surface tension.	**W**	G
d.	Has a lower heat capacity.	**W**	G
e.	Has a higher heat of vaporization.	**W**	G

Objective 2: List the major sources and effects of important water pollutants.

Oxygen-demanding wastes come from agriculture and industry, and include domestic sewage, animal manure, decaying plants and animals, and industrial wastes. The bacteria that break down these wastes consume oxygen. If the oxygen level drops too greatly, aquatic life will suffocate. Other effects include property damage (such as loss of real estate value) and aesthetic insult due to odor and taste being affected.

Disease-causing agents (bacteria, viruses, parasites) come from human and animal wastes. The major effect is of course damage to human and animal health.

Inorganic chemicals: Acids in water come from acid deposition, mining (particularly drainage from abandoned mines) and other industries. This acidity can cause not only aesthetic insult, but also property damage (because of corrosion), damage to plant and animal life, and major ecosystem disruption (particularly in the case of acid deposition over large areas).

Salts in water come from mining and industrial wastes, and irrigation (as water dissolves salts from the soil it irrigates). When irrigation water evaporates and leaves the salts behind, cropland is damaged and is sometimes unusable.

Toxic metals, such as mercury and cadmium, come from mining and industrial wastes. These can cause very serious damage to human health. It is important to keep in mind that most mercury in water, at least in the oceans, is there because of natural emissions from the earth.

Synthetic organic chemicals and oil include pesticides, detergents, industrial wastes, and oil spills. Large oil spills are only part of the problem, other sources being engine oil and other lubricating oils disposed of improperly. Effects include odor, taste, damage to plant and animal life, damage to human health, and in some cases major ecosystem disruption.

Plant nutrients include phosphates and nitrates from fertilizers, animal wastes, detergents, and industrial wastes. These pollutants cause accelerated plant growth (including algae, moss, and weeds), followed by excessive decay (a process called eutrophication). This causes aesthetic insult, property damage, damage to plant and animal life, and major ecosystem disruption.

Sediments come from soil erosion due to natural runoff, agriculture, mining, and construction. Accumulation of sediment causes aesthetic insult, property damage, and major ecosystem disruption.

Radioactive substances are found naturally in water, but may increase due to use of nuclear power, and also because many radioactive materials are used in medicine, industry, and research. The major threat is to human health.

Heat is produced by industry and electric power production. It causes damage to plant and animal life because less oxygen is present in warm water. It has the potential to cause major ecosystem disruption.

Drill:

1. Match the water pollutant with its major sources. (Sources can be used more than once, and more than one source may match up with a pollutant).

a. Oxygen-demanding wastes	1. Medicine and research
b. Disease-causing agents	2. Electric power production
c. Acids	3. Animal wastes
d. Salts	4. Mining
e. Synthetic organic chemicals	5. Acid deposition
f. Plant nutrients	6. Industrial wastes
g. Sediments	7. Irrigation
h. Radioactive substances	8. Soil erosion
i. Heat	9. Detergents
j. Toxic metals	

2. Match the water pollutant with its major effects.

a. Oxygen-demanding wastes	1. Damage to human health
b. Disease-causing agents	2. Major ecosystem disruption
c. Acids	3. Property damage
d. Salts	4. Damage to plant and animal life
e. Synthetic organic chemicals	5. Aesthetic insult
f. Plant nutrients	
g. Sediments	
h. Radioactive substances	
i. Heat	
j. Toxic metals	

Objective 3: Explain the differing solubilities of oxygen and salts in water, and how solubility changes as a function of temperature.

Solubility is affected by the nature of the particles of the solute and solvent. Review the "like dissolves like" principle in Section 4.7. Since water is polar and oxygen is nonpolar, the solubility of oxygen in water is low (only 9 mg per liter, or 9 ppm, at about room temperature). Since salts are ionic, many of them dissolve in much greater quantity. (There are many insoluble salts, however, that are exceptions to the "like dissolves like" principle).

Temperature also affects solubility. The solubility of gases in water decreases as the temperature of the water increases. (Notice the bubbles that form when a pan containing water is heated on the stove). Energy is released when gas molecules dissolve and are attracted to water molecules.

$$O_2 + H_2O \rightleftharpoons \text{solution} + \text{energy}$$

A review of Section 7.4 will show that adding energy will shift the equilibrium to the left, removing oxygen (or other gases) from solution.

Most salts absorb energy when they dissolve in water, so their solubility increases as the temperature increases, as predicted by LeChatelier's Principle.

$$\text{salts} + \text{water} + \text{energy} \rightleftharpoons \text{solution}$$

Some salts are exceptions, however. The hard water deposits that form inside hot water heaters are salts that are less soluble at higher temperatures.

Drill:

1. Why do carbonated drinks (that contain dissolved CO_2) go flat when they get warm?
2. Why do cold-water fish die when placed in warm water?
3. Table salt is more soluble in hot water than in cold water. Is this what you would expect from a salt?

Objective 4: Distinguish between dissolved oxygen (DO) and biochemical oxygen demand (BOD), and know how they relate to water quality.

Dissolved oxygen (DO) refers to the actual amount of oxygen dissolved in water, usually expressed in ppm. At 20 ^{o}C the maximum DO is 9 ppm, which is the limit of oxygen's solubility at that temperature.

Biochemical oxygen demand (BOD) is the amount of dissolved oxygen used up as water sits for 5 days at 20 ^{o}C. If the BOD of a water sample is 5 ppm, its DO will drop from 9 ppm to 4 ppm in 5 days. BOD results from pollution which reacts with dissolved oxygen.

Clean, unpolluted water has high DO and low BOD, while polluted water has low DO and high BOD. Refer to Figure 13.11.

Drill:

1. A water sample has a DO content of 9 ppm, which after five days has dropped to 7 ppm. What is the BOD, and

how would you rate the water quality?
2. Under what conditions would DO drop even when no pollution is present?

Objective 5: Distinguish between aerobic and anaerobic decay.

Aerobic decay occurs when dissolved oxygen is high. There is sufficient oxygen for bacteria to decompose materials present in water to CO_2, NO_3^-, SO_4^{2-}, PO_4^{3-}, and other oxygen-containing products. These products are not offensive or toxic (in reasonable amounts).

Anaerobic decay occurs when dissolved oxygen is low. The decay products generally contain little oxygen and often have offensive odors or are poisonous. Examples are CH_4, NH_3, H_2S, and PH_3.

Drill:
1. How is the odor and quality of water related to its dissolved oxygen content?
2. A body of water with a high BOD level is often smelly and offensive. Why?

Objective 6: Explain how rivers and lakes differ in their pollution problems.

River pollution usually is less serious than lake pollution. The pollutants are diluted because of a constant supply of cleaner water coming downstream. Also, moving water permits oxygen to dissolve more rapidly. DO levels recover and return to normal fairly quickly.

Lake pollution is more likely to be serious and long-term. There is little flow, and pollutants remain near the point of entry. Oxygen dissolves slowly, and recovery is slow or nonexistent. Eutrophication is a threat when the pollutants include nutrients such as nitrates and phosphates.

Drill:
1. Why is eutrophication not a problem in rivers?
2. How does the rate at which oxygen dissolves affect the quality of water?

Objective 7: Distinguish among primary, secondary, and tertiary sewage treatments.

Primary sewage treatment is a mechanical process that involves filtering and settling. Sometimes chemicals are added which speed up the settling process by forming additional solids (a process called flocculation). This removes most solid material but only part of the oxygen-demanding wastes. If this is the only treatment used, a disinfectant (usually chlorine) is added to kill harmful bacteria.

Secondary sewage treatment is a biological process in which the aerobic decay of sewage is accelerated. A common method is to inject air or oxygen into the sewage, creating activated sludge. With extra oxygen, the bacteria degrade the sewage more rapidly. Even secondary treatment leaves many pollutants, and chlorine is used as a disinfectant before the water leaves the treatment plant.

Tertiary sewage treatment is a series of specialized chemical and physical processes that remove remaining pollutants. It is not often used because of the expense. Examples are removal of phosphate by precipitation, removal of organic compounds by filtration with activated carbon, and removal of salts by reverse osmosis.

Drill:

1. Match the type of sewage treatment with the kind of process used.

a.	primary	1.	Biological
b.	secondary	2.	Chemical and physical
c.	tertiary	3.	Mechanical

2. What does flocculation do in primary sewage treatment?
3. What is activated sludge?
4. What is activated carbon used for?

Self-Test

---

1. Which would have a greater heat capacity, water or salad oil?
2. To remove the following pollutants, would it require Primary, Secondary, or Tertiary sewage treatment?

a)	sediment	P	S	T
b)	organic wastes	P	S	T
c)	phosphates	P	S	T
d)	oxygen-demanding wastes	P	S	T

3. Which of the following cannot pollute water?
   a) heat b) salt c) acid d) none of them
4. Choose the liquid with the greatest heat of vaporization:
   a) gasoline b) alcohol c) mercury d) water
5. Which one of the following pollutants does not affect the biological oxygen demand in water?
   a) salt b) sewage c) manure d) dead plants
6. How do fish get the oxygen they need underwater?
7. What is unusual about the density of ice compared to that of liquid water?

8. Does oxygen become more or less soluble in water when the temperature is raised?
9. Are Epsom salts more soluble in hot water or cold?
10. When water contains a lot of organic waste, does it decay aerobically or anaerobically?
11. Tell one effect of sediment as a water pollutant.
12. Does eutrophication occur more often in rivers or lakes?

Answers

Objective 1:

1. There is no hydrogen bonding in H_2S, so the attractive forces are weaker.
2. CO_2 molecules are linear and nonpolar, so the attractive forces are weaker.
3. No. It is unlikely that the molecules would crystallize in a hexagonal shape.
4. a) W; b) W; c) G; d) G; e) W

Objective 2:

1. a) 3,6; b) 3; c) 4,5,6; d) 4,6,7; e) 6,9; f) 3,6,9; g) 4,8; h) 1,2 ; i) 2; j) 4,6
2. a) 3,4,5; b) 1,4; c) 2,3,4,5; d) 3; e) 1,2,4,5; f) 2,3,4,5; g) 2,3,5; h) 1; i) 2,4; j) 1

Objective 3:

1. CO_2 is less soluble in warm water.
2. There is not enough oxygen in the water for them to survive.
3. Yes.

Objective 4:

1. The BOD is 2 ppm, and the water is slightly polluted.
2. When the water is heated.

Objective 5:

1. High DO levels indicate aerobic decay, which produces unoffensive products.
2. High BOD levels mean low DO levels, which indicates anaerobic decay and offensive products.

Objective 6:

1. The dilution of pollutants and addition of more oxygen prevents the condition from developing.
2. BOD causes DO to drop over time. If more oxygen dissolves, it helps to increase DO even when BOD is temporarily high.

Objective 7:

1. a) 3; b) 1; c) 2
2. It speeds up the settling process by which solids are removed from the sewage.
3. Sewage into which air or oxygen has been injected to speed up aerobic decay.
4. To filter out organic compounds.

Self-Test:

1) water; 2a) P; 2b) T; 2c) T; 2d) S; 3) d; 4) d; 5) a; 6) they "breathe" the dissolved oxygen; 7) it's less dense; 8) less; 9) hot; 10) anaerobically; 11) ecosystem disruption, property damage, or aesthetic insult; 12) lakes

Evaluation:

If you missed more than one question on the self-test in any of the following groups, you need to review the section indicated:

Question Groups	Section
1, 4, 7	13.1
3, 8, 9, 11	13.3 and 13.5
5, 6, 10, 12	13.4
2a, b, c, d	13.7

Chapter 14

Laundry Products: Getting the Dirt Out

Outline

I. Surface-Active Agents: Removing Dirt and Grease
  A. Rinsing is Not Enough
  B. Surfactants and Solubility
    1. Like dissolves like
    2. Solubilities of organic functional groups
  C. Surfactants: Molecules with Split Personalities
    1. Water-soluble, oil-repellent head
    2. Oil-soluble, water-repellent tail
  D. Surface Tension
    1. Tendency for liquids to have as little surface area as possible
    2. Surfactant-water mixtures have reduced surface area
  E. Wetting
    1. Tendency for liquid to cover a surface
    2. Surfactant-water mixtures wet better
  F. Foaming
    1. Formation of bubbles or suds
    2. Surfactant-water mixtures foam more
  G. Detergent Action
    1. Covering of oil-droplets with surfactant molecules
    2. Allows otherwise-insolubles substances to be mixed with water

II. Soap and Hard Water: Ring Around the Collar
  A. The Eariest Surfactant
    1. Soap
    2. Salts of fatty acids
  B. The Hard Water Problem
    1. Hard water--water that contains Ca^{2+}, Mg^{2+}, or Fe^{2+} ions
    2. Soap combines with these ions to form an insoluble soap curd
    3. Modern washing machines cannot remove soap curd
  C. Soft Water
    1. Natural soft water
    2. Water softeners

III. Synthetic Laundry Detergents: Replacements for Soap
  A. Synthetic Detergents
    1. Major ingredients
      a. Surfactants
      b. Builders
      c. Others
    2. Kinds of surfactants
  B. Anionic Surfactants

1. Water-soluble head has negative charge
2. Sodium alkylbenzene sulfonates
   a. Effective as soap
   b. Does not form insoluble curd in hard water

C. Cationic Surfactants
1. Water-soluble head has positive charge
2. Used in fabric softeners

D. Nonionic Surfactants
1. Water-soluble head has no charge but is very polar
2. Pros and cons
   a. Avoids hard water reactions
   b. Does not keep soil well-suspended in rinse

E. Biodegradability
1. Bacterial decomposition of oil-soluble tails
   a. Branched tail not degradable
   b. Linear tail biodegradable
2. All detergents must be biodegradable by law

IV. Builders: Better Cleaning

A. Detergent Teamwork

B. Phosphates
1. Sodium tripolyphosphate (STPP)
2. Detergent actions
   a. Surrounds and immobilizes mineral ions
   b. Keeps pH in alkaline range
   c. Adds to cleaning power of surfactant

C. Water Pollution
1. Eutrophication
2. Bans and limitations of phosphates

D. Phosphate Substitutes
1. Now used
   a. Sodium citrate
   b. Zeolites
2. Possible future substitutes

V. Other Detergent Ingredients: Making Detergents Better

A. Bleaches
1. Oxidizing agents that decolorize but do not remove stains
2. Sodium perborate commonly found in detergents

B. Auxiliary Builders (borax)

C. Brighteners
1. Bluing--blue dye to cancel yellowing
2. Optical brighteners--organic compounds that change invisible UV light to visible blue

D. Antiredeposition agents

E. Enzymes

F. Miscellaneous Ingredients

Objectives

After you read and study the chapter [and the sections in brackets], you should be able to:

[C] 1. Identify, by their formulas, molecules that act as surfactants in water. [Section 14.1; Questions 1, 2, 3, 4]

[C] 2. Predict how the properties of surfactant-water mixtures differ from those of aqueous solutions. [Section 14.1; Questions 5, 6, 7, 8, 9]

[C] 3. Distinguish hard water from soft water given the identity of the solutes dissolved in it or its reactions with soaps and detergents. [Section 14.2; Questions 10, 11, 12, 13]

[C] 4. Identify and tell the function of the major ingredients in a laundry detergent. [Sections 14.3, 14.4, 14.5; Questions 14, 15, 16, 17, 18, 19, 20, 21, 22]

Practice

Objective 1: Identify, by their formulas, molecules that act as surfactants in water.

A surfactant must have both of the following molecular structures: a water-soluble head consisting of an ionic group (including salts of carboxylic and sulfonic acids) or a long, uncharged organic group containing more than 12 oxygens (see Table 14.1, p. 383) and an oil-soluble tail consisting of a hydrocarbon group of 12 to 20 carbons.

Examples:

1. $CH_3(CH_2)_{12}CH_2\text{-}\overset{O}{\overset{\|}{C}}\text{-}O^-Na^+$. SURFACTANT. The hydrocarbon group contains 14 carbons (which is between 12 and 20), and the other group is the ionic salt of a carboxylic acid.

2. $Na^+O^-\text{-}\underset{O}{\overset{O}{\overset{\|}{\underset{\|}{S}}}}\text{-}O\text{-}CH_2(CH_2)_3CH_3$. NOT A SURFACTANT. Only 8 C's.

3. $CH_2CH_2\text{-}O\text{-}CH_2CH_2\text{-}O\text{-}CH_2(CH_2)_{30}CH_3$. NOT A SURFACTANT. Head contains fewer than 12 oxygens; tail has more than 20 carbons.

Drill:

1. For each of the following molecules, tell whether it is a Surfactant or Not:

   a) $H\text{-}O\text{-}(CH_2CH_2\text{-}O)_{12}\text{-}CH_2(CH_2)_{10}CH_3$ S N

   b) $Na^+O^-\text{-}\overset{O}{\overset{\|}{C}}\text{-}CH_2(CH_2)_{19}CH_3$ S N

   c) $CH_2\text{-}\underset{CH_3}{\underset{|}{CH}}\text{-}(CH_2\text{-}\underset{CH_3}{\underset{|}{CH}})_4CH_2\text{-}O\text{-}\underset{O}{\overset{O}{\overset{\|}{\underset{\|}{S}}}}\text{-}O^-K^+$ S N

   d) $CH_3(CH_2)_{12}\text{-}O\text{-}Cl$ S N

Objective 2: Predict how the properties of surfactant-water mixtures differ from those of aqueous solutions.

Surfactant-water mixtures have four properties: 1) reduced surface tension, 2) increased wetting, 3) foam production, and 4) detergent action.

Examples:

1. Forming beads and droplets on a surface. Aqueous solutions, with more surface tension and less tendency to wet, do it more than surfactant-water mixtures.
2. Frothing. Surfactant-water mixtures have more tendency to foam.

Drill:

1. Indicate whether a Surfactant-water mixture or an Aqueous solution will have the more pronounced effect:

a)	specks of paper float	S	A
b)	oil spots dissolve	S	A
c)	suds form	S	A
d)	a pane of glass gets coated	S	A
e)	tiny crack in container leaks slower	S	A
f)	dirty dishes come clean	S	A
g)	flatter, less rounded surface in a tube	S	A

Objective 3: Distinguish hard water from soft water given the identity of the solutes dissolved in it or its reactions with soaps and detergents.

Hard water has Ca^{2+}, Mg^{2+}, or Fe^{2+} ions dissolved in it. It forms insoluble soap curd, and it diminishes a detergent's cleaning power. In fact, all surfactant attributes are lessened.

Soft water lacks these properties.

Examples:

1. Contains Na^{+} ions. SOFT WATER. Only the three specified ions commonly cause water to be hard.
2. Bathtub ring forms. HARD WATER. Soap curd.

Drill:

1. Tell whether the following happens in Soft water or Hard water:

a)	clothes get cleaned better	S	H
b)	detergents must "work harder"	S	H
c)	spots appear on glassware	S	H
d)	soapy film forms	S	H
e)	more suds	S	H
f)	washed dishes come out less greasy	S	H

Objective 4: Identify and tell the function of the major ingredients in a laundry detergent.

Ingredients in laundry detergents can be classified into three categories: 1) surfactants (anionic, cationic, or nonionic) which do the actual cleaning; 2) builders which soften the water, maintain alkaline pH, and otherwise assist the surfactant; and 3) other ingredients which include bleach, brighteners, enzymes, antiredeposition agents, washer protection agents, manufacturing aids, colors, and fragrances.

Laundry detergent packages have some of the best ingredient labels of any product. Generally each of the above categories is identified clearly.

Example:

1. Quote from a Procter and Gamble product: "BIZ combines two types of enzyme stain-removers, an agent to soften water and improve cleaning (sodium tripolyphosphate), plus a fabric-safe oxygen bleach (sodium perborate); also a fabric whitener, surfactant, quality control agents, blueing, and perfume in small quantities." It would be hard to classify these incorrectly. BIZ contains enzymes, a builder, a bleach, a brightener, a surfactant, manufacturing aids, another brightener, and fragrances, respectively.

Drill:

1. Classify the ingredients on the label of Lever Brothers' All:

   "Cleaning agents (nonionic surfactants, soap), water softener (sodium carbonate); fabric brightening agent, perfume, anti-redeposition agent, colorant, washer protection agent (sodium silicate), and processing aid."

Self-Test

---

1. Will $CH_3(CH_2)_{12}\overset{O}{\overset{\|}{C}}-O^-Na^+$ act as a surfactant?
2. When Cl^- ions are dissolved in water, the solution is called hard water. True or false?
3. The principle cleaning ingredient in any laundry detergent is known as a ______________.
4. A carefully-placed sewing needle will float better on the surface of pure water or a surfactant-water mixture?
5. Why isn't hexadecane, $CH_3(CH_2)_{14}CH_3$, a surfactant?
6. Laundry bleach
   a) oxidizes stains b) decolorizes stains
   c) both of these d) neither of these

7. A builder in a laundry detergent
   a) adds body to clothes   b) softens the water
   c) lifts out the dirt   d) keeps dirt from redepositing
8. Bath tub rings are more likely to occur when bath water is soft or hard?
9. True or false: modern laundry surfactants are not bothered by hard water.
10. Is the surface tension of water increased or decreased when a surfactant is added?

Answers

Objective 1:
1. a) S; b) N; c) S; d) N

Objective 2:
1. a) A; b) S; c) S; d) S; e) A; f) S; g) A

Objective 3:
1. a) S; b) H; c) H; d) H; e) S; f) S

Objective 4:
1. surfactants, builder, brightener, fragrance, antiredeposition agent, color, washer protection agent, manufacturing aid

Self-Test:

1) yes; 2) false; 3) surfactant; 4) pure water; 5) no water-soluble head; 6) c; 7) b; 8) hard; 9) false; 10) decreased

Evaluation:
If you missed more than one question on the self-test in any of the following groups, you need to review the section indicated:

Question Groups	Section
1, 4, 5, 10	14.1
2, 8, 9	14.2
3, 6, 7	14.3 - 14.5

Chapter 15

Synthetic Polymer Products: Making Plastics, Clothing, and Tires

---

Outline

I. Polymers: Really Big Molecules
   - A. Natural and Synthetic Polymers
   - B. A Plastics World: Some Pros and Cons
   - C. Types of Polymerization Reactions
     1. Addition polymerization
     2. Condensation polymerization
   - D. Structural Variety in Polymers

II. Addition Polymers: Polyethylenes, Vinyls, and Synthetic Fibers
   - A. Polyethylene
   - B. Vinyl Addition Polymers: The Effects of Side Groups
     1. Polypropylene
     2. Polyvinylchloride (PVC)
     3. Polyvinylidene chloride (Saran)
     4. Polytetrafluoroethylene (Teflon)
     5. Polystyrene (Styrene, Styrofoam)
     6. Polyacrylonitriles (Orlon, Acrilan, Creslan)
     7. Polyvinylpyrrolidene (PVP)
     8. Polymethyl methacrylate (Lucite, Plexiglas)

III. Condensation Polymers: Dacron, Nylon, and Thermosetting Polymers
   - A. Polyesters: Dacron
   - B. Polyamides: Nylons
   - C. Thermosetting Polymers

IV. Other Polymers: Rubber and Silicones
   - A. Elastomers
   - B. Polymers and paints
   - C. Silicones

Objectives

After you read and study the chapter [and the sections in brackets], you should be able to:

[C] 1. Explain what polymers are, why they are so common, what two main types of polymerization reactions occur, and how polymers vary in structure. [Section 15.1; Questions 1, 2, 3]

[C] 2. Identify the structures and uses of polyethylene and vinyl addition polymers. [Section 15.2; Questions 5, 6, 7, 8, 9, 10]

[C] 3. Identify the structures and uses of various condensation polymers. [Section 15.3; Questions 4, 11, 12, 13]

[C] 4. Explain how synthetic rubber polymers and silicone polymers are used, and identify their general structures. [Section 15.4; Questions 14, 15, 16]

Practice

Objective 1: Explain what polymers are, why they are so common, the two main types of polymerization reactions, and how polymers vary in structure.

Polymers are very large molecules (macromolecules) made from many small, repeating molecular units called monomers.

Polymers are very common because they have been relatively inexpensive to produce from fossil fuels (mostly petroleum); in addition, they can be tailored to have a wide variety of desired properties for particular uses.

The two main types of polymerization are addition polymerization and condensation polymerization. Addition polymerization occurs with monomers containing one or more carbon-to-carbon double bonds and proceeds without the removal of any atoms in the monomer units. In condensation polymerization, the monomers join together by splitting out a small molecule, usually H_2O.

Polymer chains may be linear (continuous) or branched. Cross-linked polymers have branches that join one polymer chain to another.

Examples:

1. The addition polymer polyethylene is made from the monomer ethylene; no atoms from the ethylene are removed when the monomer units join together.
2. Branched-chain polymers are softer, less dense, and less rigid than linear polymers, which can pack closer together than the branched-chain polymers.
3. Natural polymers such as proteins, starch, RNA, and DNA are macromolecules that are made from hundreds, or even thousands, of monomer units joined together.

Drill:

1. Polymer molecules that are joined to each other by covalent bonds are said to be a) condensation polymers, b) monomers, c) cross-linked, d) branched, e) linear
2. The raw material for most synthetic organic polymers is a) natural gas, b) wood, c) coal, d) cotton, e) petroleum

3. Polymers formed by the removal of a molecule of water when the monomer units join is a) a condensation polymer, b) an addition polymer, c) cross-linked, d) branched, e) linear

Objective 2: Identify the structures and uses of polyethylene and vinyl addition polymers.

Polyethylene is the most widely used plastic on the market today. Its uses include building materials, electrical insulation, toys, squeeze bottles, and plastic bags. It is made from the monomer ethylene, which has the structure $CH_2=CH_2$.

The vinyl addition polymers include polypropylene, polyvinylchloride (PVC), polyvinylidene chloride (Saran), polystyrene (Styrene, Styrofoam), polytetrafluoroethylene (Teflon), polyacrylonitrile (Orlon, Acrilan, Creslan), polymethyl methacrylate (Lucite, Plexiglas), and polyvinylpyrrolidene (PVP). All of these are addition polymers made from monomers consisting of ethylene with one or more of the hydrogen atoms replaced by a substituent atom or group.

See Table 15.1 in the text to review the structures of the monomer units and the polymers for all these addition polymers. Also review the uses for these polymers.

Examples:

1. Teflon is used for electrical insulation, gasket bearings, coating nonstick cookware, and tubing for corrosive chemicals.
2. $CH_2=CH-$ is called a vinyl group, so $CH_2=CHCl$ is vinyl chloride, which is the monomer for polyvinylchloride (PVC).
3. Polymethyl methacrylate is used for latex paints, windows, signs, and molded items such as combs.

Drill:

1. Write the name of the polymer formed from each of the following monomer units:

   a) $CH_2=CH_2$ b) $CF_2=CF_2$ c) $CH_2=CH(CH_3)$ d) $CH_2=CH(C_6H_5)$

2. The polymer from which monomer in Question 1 is used for
   a) insulating food and drink coolers
   b) indoor-outdoor carpeting and fishing nets
3. Write structural formulas to show the repeating units for each of the polymers formed from the monomers in Question 1.

Objective 3: Identify the structures and uses of various condensation polymers.

Condensation polymers include polyesters, polyamides, and various thermosetting polymers.

The polyesters, such as Dacron, are made from two monomers; one contains two carboxylic acid functional groups and the other contains two alcohol functional groups. The acid group in one monomer reacts with the alcohol group in the other monomer to eliminate water and produce an ester linkage; repeated reactions of this type produce a polymer containing many ester linkages. Polyesters are used in clothing and fabrics (especially Dacron), recording tape and computer diskettes (Mylar), and break-proof windows (Lexan).

The polyamides, such as nylon, are made from two monomers--one containing two carboxylic acid groups and the other containing two amine groups. The acid and amine groups react, eliminating water, to form amide groups; the polymer thus has many repeating amide groups. Nylon is used for a wide variety of fabrics, small machine parts, and other products.

Thermosetting polymers are polymers that become permanently hard and rigid once they have melted. Thermosetting polymers that are condensation polymers include Bakelite, Melmac, polyurethanes, and epoxy polymers.

Examples:

1. Epoxy polymers, unlike the other condensation polymers listed above, are formed by the splitting out of HCl (not H_2O) when the monomer units join. These polymers are often used to fasten two surfaces together.
2. Nylon-66 is made from monomer units that both contain 6 carbon atoms. Other types of nylon are made from monomer units with a different number of carbon atoms.
3. Dacron is made from the monomer ethylene glycol and terephthalic acid, which have the following structures:

$HO-CH_2-CH_2-OH$

ethylene glycol

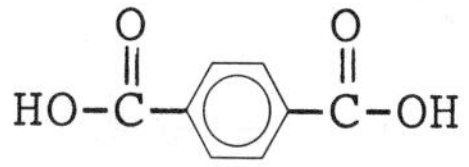

$HO-\overset{O}{\overset{\|}{C}}-C_6H_4-\overset{O}{\overset{\|}{C}}-OH$

terephthalic acid

Drill:

1. A polymer with a repeating structure of $-\overset{O}{\overset{\|}{C}}-\underset{H}{\underset{|}{N}}-$ is a(n)

   a) polyester, b) polyamide, c) epoxy polymer, d) addition polymer, e) phenol-formaldehyde polymer
2. A polymer that becomes hard and rigid after melting is called a) thermoplastic, b) branched, c) elastomer, d) thermosetting, e) vinyl polymer
3. A specific example of the type of polymer in Question 1 above is a) nylon, b) Dacron, c) Bakelite, d) Melmac, e) Teflon

Objective 4: Explain how synthetic rubber polymers and silicone polymers are used, and identify their general structures.

Polymers with the type of elasticity that occurs in rubber are called elastomers. Synthetic rubbers are addition polymers that usually are made from a monomer unit such as butadiene:

```
H   H H   H
 \  | |  /
  C=C-C=C
 /       \
H         H
```

Replacing a hydrogen atom on an interior carbon with chlorine gives the monomer chloroprene that is used to make neoprene rubber. Replacing such a hydrogen atom with a methyl (CH_3-) group makes the monomer isoprene, which is used to make natural rubber.

Vulcanization is a process to make sulfur-to-sulfur cross links in rubber; this makes the rubber more durable.

Silicone polymers are long chains of alternating silicon and oxygen atoms, with each silicon additionally bonded to two other atoms. The general structure is:

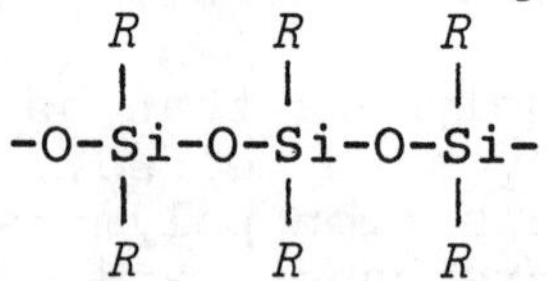

where *R* is an organic group or an oxygen atom that cross-links the polymer.

Examples:

1. The most important synthetic rubber, SBR (styrene-butadiene rubber), is made from butadiene and styrene monomer units.
2. Silicone polymers are used for car waxes, lubricants at high temperatures, surgical implants for cosmetic purposes, and waterproof coatings.
3. Using vulcanized rubber in automobile tires makes the tires more durable.

Drill:

1. The element most similar to carbon (see the periodic table) that can form polymers is a) N, b) Si, c) H, d) P, e) Ca
2. Vulcanization is a process for treating rubber with a) latex, b) steel, c) nylon, d) dyes, e) sulfur
3. Synthetic and natural rubbers are a) addition polymers, b) thermosetting polymers, c) elastomers, d) epoxy polymers, e) condensation polymers

Self-Test

Fill in the blanks appropriately:

1. _______ term for small molecules from which polymers are made
2. _______ specific example of a polyester
3. _______ simplest and most widely used addition polymer
4. _______ type of polymerization reaction in which no atoms are split out from the molecules that join together
5. _______ type of polymer that becomes rigid after melting
6. _______ polymer for which butadiene is a monomer
7. _______ type of polymer not based on carbon
8. _______ monomer unit for Styrofoam
9. _______ element, besides carbon and hydrogen, present in neoprene rubber
10. _______ elements, besides carbon and hydrogen, present in nylon
11. _______ effect (increase or decrease) of branching on the density of a polymer
12. _______ trade name of an addition polymer that contains fluorine

Answers

<u>Objective 1:</u>

1. c
2. e
3. a

<u>Objective 2:</u>

1. a) polyethylene b) polytetrafluoroethylene (Teflon) c) polypropylene d) polystyrene (Styrene, Styrofoam)
2. a) polystyrene (d) b) polypropylene (c)
3. a) ---CH₂—CH₂--- (C—C with H, H above and H, H below) b) ---C—C--- (F, F above and F, F below) c) ---C—C--- (H, H above; H, CH_3 below) d) ---C—C--- (H, H above; H and a benzene ring below)

<u>Objective 3:</u>

1. b
2. d
3. a

Objective 4:

1. b
2. e
3. a and c

Self-Test:

1) monomer; 2) Dacron (Mylar, Lexan); 3) polyethylene; 4) addition polymerization; 5) thermosetting polymer; 6) rubber (elastomer); 7) silicone polymers; 8) styrene; 9) Cl; 10) O and N; 11) decrease; 12) Teflon

Evaluation:

If you missed more than one question on the self-test in any of the following groups, you need to review the section indicated:

Question Groups	Section
1, 4, 11	15.1
3, 8, 12	15.2
2, 5, 10	15.3
6, 7, 9	15.4

Chapter 16

Personal Products:
Taking Care of Our Teeth, Skin, and Hair

Outline

I. Teeth, Tooth Decay, and Toothpastes: Clean and Healthy
  A. What Are Teeth?
    1. Parts of teeth
      a. Dentin
      b. Enamel
    2. Hydroxyapatite material
    3. Mineralization and demineralization
  B. Tooth Decay and Gum Deterioration
    1. Despite fluorides, these are the most prevalent diseases in U.S. after common cold
    2. Results from acid-caused demineralization
    3. Acid carriers
      a. Plaque
      b. Tartar
    4. Sucrose is the chief culprit
  C. Using Fluorides to Combat Tooth Decay
    1. Forms fluoroapatite
      a. Fluoride ions replace hydroxide
      b. Much less affinity for hydronium ions
    2. Inhibits bacterial enzymes
    3. Available in several forms
  D. What's in Toothpaste?
    1. Abrasives for cleaning
    2. Sodium pyrophosphate for tartar control
    3. Antibacterial to combat bad breath
    4. Fluoride to prevent tooth decay
      a. NaF
      b. SnF_2
      c. Na_2PO_3F

II. Skin Care Products: Clear and Supple
  A. Your Skin
    1. Stratum corneum
    2. Sebaceous glands
    3. Keratin
  B. Skin Cleansers
    1. Rinsing dirt off with soap or detergents
      a. Surfactant properties of soap
      b. Fatty acids added for mildness
    2. Tissuing dirt off with oils as solvents
    3. Dusting dirt off with powders as absorbers
  C. Moisturizers
    1. Keratin fibers need moisture to remain smooth and supple
    2. Humectants to draw moisture from air

        a. Water attracting properties of surfactant head
        b. Alcohols with multiple hydroxyl groups
    3. Emollients to prevent escape of moisture from skin
        a. Water repelling properties of surfactant tail
        b. Oils of all varieties
  D. Acne
    1. Clogged ducts from sebaceous glands
    2. Corrected by skin irritants
    3. Benzoyl peroxide and free radicals

III. Shampoos and Conditioners: Shiny and Manageable
  A. Hair Care
    1. Made of lifeless keratin
    2. Care of hair is up to you
  B. Shampoos
    1. Mild surfactants to remove dirt but leave optimum amount of sebum
    2. Acidifiers to give hair maximum wet strength and prevent damage
    3. Other ingredients to satisfy uneducated consumers
  C. Conditioners
    1. Positively-charged amines to prevent fly-aways
    2. Hydrocarbon attachments to lubricate combing and prevent tangles
    3. Protein fragments to repair damage
    4. Other ingredients of little consequence
    5. Conditioners, separate from shampoos, not always necessary
  D. Dandruff
    1. Abnormal flaking of scalp cells
    2. Methods of control
        a. Slowing runaway migration of scalp cells
        b. Breaking up flakes into insignificant pieces

IV. Deodorants and Antiperspirants: Dry and Confident
  A. Perspiration
    1. Eccrine sweat
        a. Body's cooling mechanism
        b. No odor problem
    2. Apocrine sweat
        a. Produced in underarms and other places
        b. Becomes smelly only when bacteria grow in it
  B. Deodorant and Antiperspirant Action
    1. Inhibit production of apocrine sweat
    2. Prevent it from reaching skin
    3. Kill offending bacteria
    4. Decompose foul-smelling substances
    5. Mask odors with perfume
  C. Antiperspirants Versus Deodorants
    1. Government labeling requirements
    2. Active ingredients

V. Sun Products: Tan and Smooth
  A. Ultraviolet Light
    1. One part of sunlight that interacts with molecules of your body

2. Beneficial action: synthesis of Vitamin D
3. Harmful action: aging of skin

B. Sunburns and Suntans
   1. Sunburn: pain from sudden high levels of UV radiation
   2. Suntan: darkening of skin from prolonged UV radiation
      a. Melanin production
      b. Absorption of UV radiation
   3. Suntans as warning signals or status symbols

C. Sunscreens
   1. Substances that absorb UV light
      a. Absorbers of all light
      b. Absorbers of dangerous UV only
   2. Sunscreen use and labeling

D. Sunless, Quick Tans
   1. Substances that form brownish complexes with skin
   2. Variable quality of resulting artificial tans

Objectives

After you read and study the chapter [and the sections in brackets], you should be able to:

[C] 1. a. Identify the main causes of tooth decay and some methods of preventing it. [Section 16.1, Questions 1, 2, 3]
   b. Identify and tell the function of the active ingredients in toothpastes [Section 16.1, Question 4]

[C] 2. Identify and tell the function of the active ingredients in skin cleansers, moisturizers, and acne medications. [Section 16.2, Questions 5, 6, 7, 8, 9]

[C] 3. Distinguish the active ingredients from the inert ingredients in shampoos, conditioners, and dandruff preparations. Tell the function of the active ingredients. [Section 16.3, Questions 10, 11]

[C] 4. Identify and tell the function of the active ingredients in antiperspirants and deodorants. [Section 16.4, Questions 12, 13, 14]

[C] 5. Identify and tell the function of the active ingredients in suntan products. [Section 16.5, Questions 15, 16, 17]

Practice

Objective 1a: Identify the main causes of tooth decay and some methods of preventing it.

Tooth decay is **caused** by the demineralization of tooth enamel, the shifting toward the right of the equilibrium

$$Ca_5(PO_4)_3OH \;=\; 5\ Ca^{2+} \;+\; 3\ PO_4^{3-} \;+\; OH^-$$

Tooth decay can be **prevented** by achieving conditions that are unfavorable to this shift in equilibrium. (Review LeChatelier's principle, if necessary: Objective 6, Chapter 7)

The main causes and preventions focus on the OH^- part of the equilibrium. Plaque and tartar produce acid which consumes the OH^- which drives the equilibrium to the right. Fluoride replaces the OH^- and circumvents the acid problem.

Examples:

1. Constantly chewing calcium-containing antacids. PREVENTS tooth decay because increased concentrations of Ca^{2+} drives the equilibrium to the left.
2. Eating candy all day long. CAUSES tooth decay because the sucrose nourishes acid-producing bacteria in the mouth. The acid, in turn, depletes the OH^- concentration and drives the equilibrium to the right.
3. Drinking a can of pop (acidic). NEGLIGIBLE effect because it does not provide teeth with a constant exposure to the acid.

Drill:

1. Tell whether the following activities Cause tooth decay, Prevent it, or have Negligble effect:

a)	daily brushing	C	P	N
b)	getting caramel stuck in teeth	C	P	N
c)	drinking fluoridated water	C	P	N
d)	chewing sugarless gum	C	P	N
e)	gargling with antiseptic	C	P	N
f)	having dentist remove tartar	C	P	N
g)	eating red meat	C	P	N
h)	breath mint always in mouth	C	P	N

Objective 1b: Identify and tell the function of the active ingredients in toothpastes.

Active ingredients include: abrasives to remove plaque and polish tooth enamel (calcium or silicon compounds);

substances that prevent tartar buildup (sodium pyrophosphate); antibacterials to fight bad breath (sodium N-lauroyl sarcosinate); fluoride to prevent tooth decay (sodium fluoride, stannous fluoride, MFP)

Examples:

1. Calcium pyrophosphate. ABRASIVE. It's a calcium compound. (Note that the tartar control ingredient is sodium pyrophosphate.)
2. Saccharin for sweetness. INERT INGREDIENT. Not on the above list.

Drill:

1. A tube of Proctor and Gamble's Crest toothpaste lists the following ingredients. Tell the function of each or label it as inert.
   a) calcium pyrophosphate
   b) sorbitol
   c) sodium fluoride
   d) sodium pyrophosphate
   e) cellulose gum
2. Colgate-Palmolive's Ultra Brite contains the following. Tell the function of each or label it as inert.
   a) sodium monofluorophosphate
   b) hydrated silica alumina
   c) titanium dioxide
   d) N-lauroyl sarcosinate
   e) sodium saccharin

Objective 2: Identify and tell the function of the active ingredients in skin cleansers, moisturizers, and acne medications.

Skin cleansers have three possible types of active ingredients: surfactants, solvents, or absorbers.
Moisturizers have either emollients, which can be any or a great variety of oils, or humectants, which are principally glycerin, propylene glycol, or sorbitol.
Acne medicines contain skin irritants. Benzoyl peroxide is a common ingredient.

Examples:

1. A moisturizing formula contains mineral oil and propylene glycol. The first is an EMOLLIENT and the second, a HUMECTANT.
2. A certain cream cleanser contains mineral oil and sodium laureth sulfate (a detergent). Here the mineral oil acts as a SOLVENT, and the detergent is a SURFACTANT.
3. "Oxy-10" acne medication contains 10% benzoyl peroxide, a SKIN IRRITANT.

Drill:

1. Identify the emollient and the humectant in both of the following moisturizer recipes:
   a) Carbomer 934, 63%; lanolin oil, 21%; propylene glycol, 5%, lanolin, 4 %, other ingredients, 7%
   b) Water, 40%; beeswax, 13%; sorbitol, 10%; palm kernal oil, 5%; other ingredients, 32%
2. Tell whether the following skin cleanser ingredients are **Sur**factants, **Sol**vents, or **Abs**orbers:

a)	corn starch	Sur	Sol	Abs
b)	alkyl benzene sulfonate	Sur	Sol	Abs
c)	talcum powder	Sur	Sol	Abs
d)	glycerin	Sur	Sol	Abs
e)	beeswax	Sur	Sol	Abs
f)	soap	Sur	Sol	Abs

3. Some acne medicines contain resorcinol as an active ingredient. What does resorcinol do to skin?

Objective 3: Distinguish the active ingredients from the inert ingredients in shampoos, conditioners, and dandruff preparations. Tell the function of the active ingredients.

Shampoos contain mild surfactants, and acidifiers as active ingredients. (Conditioning ingredients may also be present.) Foamers and thickeners do nothing for the hair.

Conditioners contain positively-charged amines, and protein fragments as their main active ingredients. Other ingredients such as oils, carbohydrate, vitamins, or botanicals may serve a minor purpose.

Dandruff preparations may contain zinc pyrithione, sulfur, salicylic acid, or selenium sulfide.

Example:

1. A shampoo contains: water, 61%; sodium lauryl sulfate, 15%; Quarternium-15 (a positive amine), 14%; hydrolized animal protein; 4%; citric acid, 2%; xanthan gum, 2%; lauramide DEA, 1%; and zinc pyrithione, 1%. These ingredients, respectively, are: inert (solvent); surfactant; conditioner; conditioner; acidifier; inert (thickener); inert (foamer); dandruff medication.

Drill:

1. Identify the following shampoo ingredients as Surfactants, Acidifiers, Conditioners, Dandruff treatment, or Inert:

a)	Hydrolyzed collagen protein	S	A	C	D	I
b)	phosphoric acid	S	A	C	D	I
c)	ammonium lauryl sulfate	S	A	C	D	I
d)	alcohol	S	A	C	D	I
e)	selenium sulfide	S	A	C	D	I

		S	A	C	D	I
f)	sodium chloride	S	A	C	D	I
g)	sodium lauryl ether sulfate	S	A	C	D	I
h)	fragrance	S	A	C	D	I
j)	colloidal sulfur	S	A	C	D	I
k)	Coco Betaine (+ amine)	S	A	C	D	I

Objective 4: Identify and tell the function of the active ingredients in antiperspirants and deodorants.

The active ingredients in nearly all commercial **antiperspirants** are aluminum or zirconium compounds. **Deodorants** have antibiotics like Neomycin or benzalkonium chloride.

Examples:

1. Aluminum zirconium tetrachlorohydrex. ANTIPERSPIRANT. Has Al and Zr; the rest doesn't matter.
2. Benzethonium chloride. DEODORANT. The "alk" part of benz**alk**onium chloride can be replaced with other syllables.
3. Ethyl alcohol. INERT.

Drill:

1. If a product lists the each of the following substances as its only active ingredient, could it be called an **A**ntiperspirant, a **D**eodorant, or **N**either?

		A	D	N
a)	aluminum chlorohydrate	A	D	N
b)	propylene glycol	A	D	N
c)	fragrance	A	D	N
d)	neomycin	A	D	N
e)	zirconium chlorhydroxide	A	D	N
f)	benzmethonium chloride	A	D	N

Objective 5: Identify and tell the function of the active ingredients in suntan products.

There are only two types of active ingredients: sunscreens (most commonly PABA and its esters, benzophenone and its derivatives) or artificial tanners (dihydroxyacetone or muconic aldehyde)

Product labels for sunscreens generally list the active ingredients only. Thus if the ingredient is not one of the two artificial tanners, it must be a sunsceen. The size of the SPF number (2-15) tells the effectiveness of the sunsreen.

Example:

1. One listed ingredient is "Padimate O". SUNSCREEN. Not dihydroxyacetone or muconic aldehyde.

Drill:

1. Tell whether each of the following active ingredients is a Sunscreen or and Artificial tanner:

a)	Benzophenone-6	S	A
b)	Muconic aldehyde	S	A
c)	Octyl dimethyl PABA	S	A
d)	Homosalate	S	A
e)	Dihydroxyacetone	S	A
f)	Glyceryl PABA	S	A

Self-Test

---

1. If citric acid is listed as a shampoo ingredient, it acts as
   a) a surfactant b) an acidifier
   c) a conditioner d) an inert ingredient
2. Stearyl alcohol, an oil-like ingredient in many skin cleansers, works as
   a) a surfactant b) an absorber
   c) a solvent d) none of these
3. An ingredient called aluminum chlorohydrate would likely be found in
   a) a skin cleanser b) a shampoo
   c) a toothpaste d) an antiperspirant
4. Which of the following would cause tooth decay?
   a) sucking on a penny b) flossing your teeth
   c) chewing caramels d) drinking distilled water
5. Identify the sunscreen among the following ointment ingredients:
   a) benzophenone-2 b) cetearyl alcohol
   c) triethanolamine d) methyl paraben
6. Which of the following would not be found in toothpastes?
   a) F_2 b) NaF c) SnF_2 d) Na_2PO_3F
7. Would avocado oil in a moisturizer be an emollient or a humectant?
8. Why are the best shampoos and conditioners acidic?
9. True or false, effective sunsceens cannot be transparent.
10. If benzalkonium chloride is the only active ingredient, what must the product be labeled as, deodorant or antiperspirant?
11. Ingredients to give a shampoo a forest green color and an herbal garden scent make the product more pleasant. Do they make it work better?
12. What function does a hydrated silica ingredient perform in a toothpaste?
13. Why is a separate conditioner usually unnecessary if you use a typical shampoo?
14. In what type of personal product is benzoyl peroxide found?
15. In what type of personal product is PABA found?
16. In what type of personal product is zirconium found?

Answers

Objective 1a:

1. a) P; b) C; c) P; d) N (neither causes nor prevents); e) P; f) P; g) N; h) C

Objective 1b:

1. a) abrasive; b) inert; c) fluoride; d) tartar control; e) inert
2. a) fluoride; b) abrasive; c) inert; d) antibacterial; e) inert

Objective 2:

1. a) emollient: lanolin oil and lanolin; humectant: propylene glycol b) emollient: palm kernel oil and beeswax; humectant: sorbitol
2. a) Abs; b) Sur; c) Abs; d) Sol; e) Sol; f) Sur
3. It must irritate the skin to encourage the sloughing of dried cells.

Objective 3:

1. a) C; b) A; c) S; d) I; e) D; f) I; g) S; h) I; j) D; k) C

Objective 4:

1. a) A; b) N; c) D (very weak); d) D; e) A; f) D

Objective 5:

1. a) S; b) A; c) S; d) S; e) A; f) S

Self-Test:

1) b; 2) c; 3) d; 4) c; 5) a; 6) a; 7) emollient; 8) hair has its maximum wet strength in acidic solution; 9) false; 10) deodorant; 11) no; 12) abrasive; 13) most shampoos contain 3 or 4 conditioners; 14) acne medicine; 15) sunscreen; 16) antiperspirant

Evaluation:

If you missed more than one question on the self-test in any of the following groups, you need to review the section indicated:

Question Groups	Section
4, 6, 12	16.1
2, 7, 14	16.2
1, 8, 11, 13	16.3
3, 10, 16	16.4
5, 9, 15	16.5

Chapter 17

Food-growing Products: Using Fertilizers and Pesticides

Outline

I. Fertilizers: Feeding the Plants
   - A. The Food Chain
   - B. Plant Nutrition
   - C. Fertilizers
   - D. Organic Fertilizers
   - E. Commercial Inorganic Fertilizers
   - F. Soil Acidity and Alkalinity

II. Pesticides: Decreasing the Competition
   - A. Pests and Pesticides
   - B. Insecticides
     1. Chlorinated hydrocarbons
     2. Organophosphates
     3. Carbamates
   - C. Herbicides
     1. Contact herbicides
     2. Systemic herbicides
     3. Soil sterilants

III. Pros and Cons of Pesticides: How Effective Are They?
   - A. The Case for Pesticides
   - B. The Case against Pesticides

IV. Alternative Methods of Pest Control: Working With Nature
   - A. Nonchemical Alternatives
   - B. Chemical Alternatives
   - C. Integrated Pest Management

Objectives

After you read and study the chapter [and the sections in brackets], you should be able to:

[C] 1. Identify the major nutrients plants need, and explain how plants obtain those nutrients. [Section 17.1; Questions 1, 2, 3, 4, 5]

[C] 2. Describe the major types of organic fertilizers and commercial inorganic fertilizers. [Section 17.1; Questions 6, 7, 9]

[C] 3. Identify the major types and properties of pesticides. [Section 17.2; Questions 8, 10, 12, 13]

[C] 4. Explain the advantages and disadvantages of using synthetic chemical pesticides. [Section 17.3; Questions 11, 14, 15, 17, 18]

[C] 5. Describe some alternative ways to control pests. [Section 17.3; Question 16]

Practice

Objective 1: Identify the major nutrients plants need, and explain how plants obtain those nutrients.

Plants obtain the nutrients they need from the soil, water, and air. Plants also need sunlight to help make carbon dioxide (CO_2) and water into glucose and oxygen by photosynthesis. Then they make glucose into other organic products they need.

Materials that plants need in relatively large amounts are called macronutrients. These elements are carbon (C), hydrogen (H), oxygen (O), nitrogen (N), phosphorus (P), potassium (K), magnesium (Mg), calcium (Ca), and sulfur (S).

Substances that plants need in only trace amounts are micronutrients. These elements include chlorine (Cl), copper (Cu), zinc (Zn), iron (Fe), manganese (Mn), boron (B), and molybdenum (Mo).

Table 17.1 in the text lists the macronutrients and micronutrients.

Examples:

1. Air and water supply the macronutrients carbon, hydrogen, and oxygen through carbon dioxide (CO_2) and water (H_2O).
2. All the micronutrients are supplied by the soil.
3. The soil has to provide relatively large amounts of N, P, K, Mg, Ca, and S if plants are to grow well.
4. Air provides large amounts of nitrogen in the form of nitrogen gas, N_2, but plants cannot use nitrogen directly in this form.

Drill:

1. Which of the following is not normally considered to be a macronutrient: a) phosphorus, b) iron, c) sulfur, d) calcium, e) all are macronutrients
2. In order to synthesize organic material, plants need to be supplied with a) CO_2, b) O_2, c) H_2, d) N_2, e) K
3. Which of the following is normally a micronutrient for plants: a) oxygen, b) calcium, c) sulfur, d) phosphorus, e) none of the above

Objective 2: Describe the major types of organic fertilizers and commercial inorganic fertilizers.

Fertilizers are used to correct soil deficiencies.

Organic fertilizers are made from plants or animals in the form of animal manure, green manure (plant stalks, wastes, and other crop residues), and compost. These fertilizers supply nitrogen, organic material, and other nutrients in a form plants can use directly.

Commercial inorganic fertilizers contain only the specific ingredients formulated by humans. The major macronutrients they provide are nitrogen, phosphorus, and potassium. A label with 3 numbers shows the content of N, P, and K, respectively.

Examples:

1. Animal manure from farm animals such as poultry, horses, and cattle is a rich source of nitrogen.
2. Commercial inorganic fertilizers often supply nitrogen in the form of ammonia (NH_3), urea ($CO(NH_2)_2$), and various nitrate (NO_3^-) and ammonium (NH_4^+) salts.
3. A sack of 25-10-5 fertilizer is richer in nitrogen and has less phosphorus and potassium than a 12-42-8 fertilizer.

Drill:

1. Compared with commercial inorganic fertilizers, organic fertilizers typically a) supply more C, b) are easier to store, c) have fewer micronutrients, d) supply more N, e) none of the above
2. The second number in a three-number label on a bag of fertilizer indicates its content of a) N, b) S, c) K, d) P, e) O
3. Ionic compounds may be used in commercial inorganic fertilizers to supply a) N, b) P, c) K, d) all of the above, e) none of the above

Objective 3: Identify the major types and properties of pesticides.

The 3 major types of insecticides are chlorinated hydrocarbons, organophosphates, and carbamates.

Chlorinated hydrocarbon pesticides contain one or more chlorine atoms. They are cheap, kill a variety of target and nontarget organisms, and remain in the environment for years.

Organophosphate pesticides contain a phosphate (PO_4) or phosphatelike group. They are much more toxic to use than chlorinated hydrocarbons and usually break down in the environment within a few weeks.

Carbamate pesticides are derivatives of carbamic acid that contain a carbon atom in an ester group bonded directly to a nitrogen. These compounds are usually less toxic than organophosphates and break down in the environment in a few days or weeks.

Herbicides are classified as contact herbicides, which kill plants within a few days after direct contact; systemic herbicides, which are absorbed by the plant and cause death by excessive growth; and soil sterilants, which kill plants by destroying soil microorganisms that they need.

Examples:

1. Organochlorine insecticides include DDT, dieldrin, endrin, lindane, chlordane, and heptachlor.
2. Organophosphate insecticides include parathion, malathion, Diazinon, Phosdrin, and TEPP.
3. Carbamate pesticides include carbaryl (Sevin), Baygon, Zectran, and Temik.
4. A common contact herbicide is atrazine; 2,4-D and 2,4,5-T are systemic herbicides; and Treflan, Dymid, Dowpon, and Sutan are soil sterilants.

Drill:

1. Which of the following will normally stay in the environment for the longest time: a) parathion, b) DDT, c) carbaryl, d) malathion, e) Baygon
2. Which of the following would be the most toxic to a farm worker applying the chemical: a) DDT, b) lindane, c) malathion, d) endrin, e) heptachlor
3. An example of a systemic herbicide is a) atrazine, b) DDT, c) carbaryl, d) 2,4-D, e) none of the above

Objective 4: Explain the advantages and disadvantages of using synthetic chemical pesticides.

Synthetic chemical pesticides have important benefits. They have greatly increased the amount of food available by decreasing the loss to insects. Pesticides also have saved many millions of lives by killing insect carriers of diseases such as malaria, typhus, and sleeping sickness. Pesticides are fairly inexpensive, safe when handled properly, and available in many types depending on the target insect.

Synthetic chemical pesticides have disadvantages. They generally are not specific, so they kill other organisms besides the target pests; those unintended victims may be predators of pests, or animals valued for other reasons. Genetic resistance often develops in target pests, making those pests even more difficult to control. Through food chains, the concentration of pesticides often increases dramatically; this is called biological magnification. Some pesticides stay in the environment a long time and thus continue to harm various animals.

Examples:

1. DDT has saved an estimated 5 million lives in the last 25 years by reducing the spread of malaria by the Anopheles mosquito.
2. Through biological magnification, the concentration of chlorinated hydrocarbons can be as much as several million times greater in fish than in the water in which they live.
3. The cotton boll weevil has become largely resistant to dieldrin and aldrin, which was formerly used to kill those pests.

Drill:

1. Certain features of pesticides have both advantages and disadvantages. List an advantage and a disadvantage for each of the following characteristics of a pesticide: a) highly toxic, b) remains in the environment a long time
2. A pesticide that is no longer used in the U.S. because of its adverse environmental effects, such as causing thin egg shells in birds and thus preventing hatching of their young, is a) parathion, b) Agent Orange, c) DDT, d) rotenone, e) none of the above

Objective 5: Describe some alternative ways to control pests.

Nonchemical alternatives include cultural control (planting and rotating crops to reduce infestations), use of natural predators, sterilizing insects, and developing plants that are genetically resistant to certain pests.

Hormones and pheromones (chemicals that transmit various messages such as sex attraction) can be used to upset normal growth and development, interfere with mating, or lure insects into traps.

Integrated pest management is a plan to use all types of control--pesticides, hormones and pheromones, and nonchemical alternatives--to best minimize insect damage to particular crops in particular locations.

Examples:

1. Traps baited with a pheromone that is a sex attractant have been used to lure male gypsy moths to their death.
2. The U.S. Department of Agriculture has released many millions of sterilized (by radiation) male screwworm flies to reduce the population of these pests along the U.S. - Mexico border.
3. Rotating crops grown in a given field from year to year reduces the chances of pest infestations.

Drill:

1. Insect reproduction can directly be reduced by a) spraying the area with the sex attractant for that insect, b) releasing sterilized males into the general population, c) spraying the area with a pesticide, d) more than one of the above, e) none of the above
2. Insects can be killed by making them develop abnormally through their exposure to certain a) insecticides, b) hormones, c) herbicides, d) pheromones, e) sterilized males
3. List advantages and disadvantages of controlling insect pests by their predators.

Self-Test

Fill in the blanks appropriately:

1. _______ plant macronutrient in Group 1A in periodic table
2. _______ example of "green manure"
3. _______ insect substance that functions as a sex attractant
4. _______ type of insecticide that typically is the most toxic
5. _______ major disease whose spread has been decreased by DDT
6. _______ plant macronutrient that is in the air in a form that cannot be used directly by plants
7. _______ two nonchemical methods of insect control
8. _______
9. _______ type of insecticide that is now used less because of its long residence time in the environment
10. _______ element in fertilizer referred to by the "6" in 30-10-6
11. _______ plant micronutrient that is a halogen
12. _______ type of pesticide that kills plants within a few days by direct contact
13. _______ term for increase in concentration of a substance (such as a pesticide) that occurs in a food chain
14. _______ type of pesticide with generally the shortest time in the environment

---

Answers

Objective 1:

1. b
2. a
3. e

Objective 2:

1. a
2. d
3. d

Objective 3:

1. b
2. c
3. d

Objective 4:

1. a) advantage - will kill target pests quickly
   disadvantages - dangerous to apply; may harm other, unintended organisms in the vicinity

   b) advantage - needs to be applied only infrequently
   disadvantages - can accumulate in the environment and do other damage; enables genetic resistance to develop
2. c

Objective 5:

1. d   2. b
3. advantages - no chemicals left in the environment; predators are a "natural" control method
   disadvantages - predator populations may become too low to be effective or too high, in which case the predators themselves become a problem

Self-Test:

1) K; 2) any type of fresh, green vegetation plowed into a field as organic fertilizer; 3) pheromone; 4) organophosphate; 5) malaria; 6) N; 7) and 8) predators, sterilization, resistant crop strains, crop rotation, intercropping; 9) chlorinated hydrocarbons; 10) K; 11) Cl; 12) contact herbicide; 13) biological magnification; 14) carbamate

Evaluation:

If you missed more than one question on the self-test in any of the following groups, you need to review the section indicated:

Question Groups	Section
1, 2, 6, 10, 11	17.1
4, 9, 12, 14	17.2
5, 9, 13	17.3
3, 7, 8	17.4

Chapter 18

Nutrients and Additives in Food Products: Eating to Stay Healthy

---

Outline

I. Human Nutrition: You Are What You Eat
   A. Metabolism: The Chemical Machinery of Life
   B. Nutrition
   C. Calories
   D. Carbohydrates
   E. Fats and Oils
   F. Proteins
   G. Obesity and Dieting
II. Vitamins and Minerals: The Little Things That Count
   A. Vitamins and Nutritional Deficiency Diseases
   B. Classes of Vitamins
      1. Water-soluble vitamins
      2. Fat-soluble vitamins
   C. Minerals
III. Food Additives: How Natural Is Natural?
   A. To Add or Not to Add: That Is the Question
   B. How Safe Are Chemical Additives?
   C. Natural Versus Synthetic Foods
IV. Some Controversial Food Additives: How Safe Is Safe?
   A. Artificial Sweeteners
   B. Flavors and Flavor Enhancers: The MSG Controversy
   C. Coloring Agents: Coal Tar Dyes
   D. Emulsifiers
   E. Preservatives: Nitrates and Nitrites
   F. Antioxidants: BHA and BHT
   G. Choices

Objectives

After you read and study the chapter [and the sections in brackets], you should be able to:

[C] 1. Identify the major nutrients you need, and explain what happens to these nutrients in your body. [Section 18.1; Questions 1, 2, 3, 4, 6, 7, 10, 11]

[C] 2. Explain what happens if you get too few or too many calories, or too little protein. [Section 18.1; Questions 5, 8, 9, 12, 13, 14]

[C] 3. Identify the key vitamins and minerals you need, and explain what happens when there is a deficiency of these nutrients. [Section 18.2; Questions 16, 17]

[C] 4. Describe the major types of food additives and how natural foods differ from processed or synthetic foods. [Section 18.3; Questions 15, 18, 19]

[C] 5. Explain some of the major issues concerning the safety of food additives. [Section 18.4; Question 20]

Practice

Objective 1: Identify the major nutrients you need, and explain what happens to these nutrients in your body.

The nutrients we need include carbohydrates, lipids, proteins, water, and various vitamins and minerals.

Through digestion the body converts carbohydrates, lipids, and proteins we eat into the simpler materials from which those nutrients are made--simple sugars (especially glucose), glycerol and fatty acids, and amino acids, respectively.

In a process called metabolism, the body then chemically converts these simple nutrients into other needed materials, or oxidizes them to produce energy. In a typical American diet, carbohydrates provide about 50-60% of the Calories, lipids (fats and oils) provide 20-40%, and proteins supply about 10-15%. Lipids, when metabolized, provide about twice as many Calories per gram (9 Cal/g) than carbohydrates and proteins.

We also need water, which is the main solvent and transportation system in the body. Vitamins participate in metabolic and other processes, and minerals are essential for structure (such as bones) and for various metabolic events.

Examples:

1. Although we also need nucleic acids (DNA and RNA) in our bodies, we don't have to obtain them in the diet because we can produce them from other materials through metabolism.
2. Through digestion, we break down proteins in the diet into their individual amino acids. Then, through metabolism, our bodies reassemble those amino acids into the proteins specified by our own DNA.
3. We have enzymes to digest starch and certain other complex carbohydrates to form simple sugars, mainly glucose, that we metabolize. But we lack an enzyme to break down cellulose, a polymer of glucose, so cellulose cannot serve directly as a caloric source for us.

Drill:

1. Digestion is a process that a) oxidizes glucose to produce energy, b) assembles amino acids into proteins, c) converts fats into fatty acids and glycerol, d) produces energy, e) none of the above
2. The fuel stored in our bodies that can provide the greatest amount of energy, when metabolized, is a) carbohydrate, b) fat, c) vitamins, d) proteins, e) minerals
3. Compared to a typical American diet, people in less developed countries often have diets with a higher percentage of their energy being provided by a) carbohydrates, b) fats, c) vitamins, d) proteins, e) minerals

Objective 2: Explain what happens if you get too few or too many calories, or too little protein.

Carbohydrates, fats, and proteins are our sources of energy.

Excessive intake of these caloric sources produces obesity, with the extra fuel being stored mostly as body fat.

People who eat too little of these caloric sources are excessively thin, have little energy, and are vulnerable to disease. They also are likely to have other nutritional deficiencies, because often they don't have enough food of any kind.

People with insufficient food typically lack protein in their diets. The proteins we eat have to supply all 8 of the essential (in the diet) amino acids that we cannot produce from other materials. Proteins that do this usually come from animal sources and are called complete proteins.

When dietary protein doesn't supply enough of all the essential amino acids, people cannot produce the proteins their bodies need for muscle, bone, metabolism, and other functions. This leads to stunted growth, possible mental impairment, and a variety of other effects.

Examples:

1. Kwashiorkor and marasmus are diseases that occur in infants and very young children who lack adequate calories and essential amino acids in their diet. Many die early as a result. The diseases, if caught in the early stages, can be cured with an adequate diet.
2. In the United States, 30-35% of middle-aged adults weigh at least 20% more than their normal, desirable body weight and are thus classified as obese.
3. Corn is a source of protein, but most strains don't have adequate amounts of the essential amino acid lysine. People who use corn as their sole major source of protein thus can develop symptoms of protein malnutrition.

Drill:

1. Explain why people who get adequate amounts of protein in their diets still can develop protein malnutrition.
2. Which of the following is not a direct source of energy (calories) in the body: a) protein, b) vitamins, c) lipids, d) carbohydrates, e) none of the above
3. Symptoms of kwashiorkor typically include all of the following except: a) poor mental development, b) susceptibility to infectious diseases, c) cancer, d) stunted growth, e) none of the above

Objective 3: Identify the key vitamins and minerals you need, and explain what happens when there is a deficiency of these nutrients.

Key vitamins and minerals needed in the diet are listed below. If a disease is clearly known to result from a deficiency of the vitamin, it is listed in parentheses.

Fat-soluble (nonpolar) vitamins: A (night blindness), D (rickets), E, and K (failure of the blood to clot rapidly).

Water-soluble (polar) vitamins: B_1 or thiamine (beriberi), B_2 or riboflavin, pantothenic acid, B_5 or niacin (pellagra), B_6 or pyridoxine, B_{12} or cobalamin (pernicious anemia), biotin, C or ascorbic acid (scurvy).

Minerals: iron (Fe), zinc (Zn), copper (Cu), iodine (I), calcium (Ca), magnesium (Mg), chlorine (Cl), chromium (Cr), potassium (K), phosphorus (P), sodium (Na), sulfur (S), manganese (Mn), and fluorine (F) are some of the elements we need as minerals (in ionic form) in our bodies.

Examples:

1. Iron is an important component of the hemoglobin in our red blood cells. A shortage of iron in the diet produces a shortage of red blood cells, a condition called anemia.
2. Centuries ago sailors discovered that taking citrus fruits on long voyages kept them from getting scurvy. Now we know this protection comes from the vitamin C in citrus fruits.
3. Vitamins A, D, E, and K can be stored in the body because they are fat-soluble (nonpolar); the water-soluble (polar) vitamins, however, are excreted readily in the urine.

Drill:

1. A deficiency of vitamin D leads to a) rickets, b) night blindness, c) beriberi, d) scurvy, e) slow blood clotting
2. An element needed in the body (in ionic form) that is in Group 11 (1B) in the periodic table is __________.

3. A deficiency of the vitamin _______ or mineral ________ results in a form of anemia.

Objective 4: Describe the major types of food additives and how natural foods differ from processed or synthetic foods.

Review the major types of food additives listed in Table 18.3 in the text. They include preservatives, antioxidants, nutrition supplements, coloring and flavoring agents, acidulants (to provide a tart taste), alkalis (to neutralize acidity), emulsifiers, stabilizers and thickeners, sequestrants (to tie up trace amounts of metal ions), and contaminants that are not deliberately added to foods.

Whether natural or synthetic, a given chemical, in pure form, has the same effect in the body. Natural foods contain substances that are known to be toxic, and synthetic or processed foods also are likely to contain trace amounts of harmful substances. Natural foods are more likely to contain trace amounts of beneficial substances not present in synthetic or processed foods. In some cases, processed foods have additives to restore nutrients lost during processing. Foods stored for lengthy periods of time often have additives to protect them against deterioration.

Examples:

1. More than 1000 substances are used as flavoring agents; they include monosodium glutamate (MSG), saccharin, and aspartame.
2. Antioxidants used to prevent undesirable oxidation of foods, particularly those containing lipids, include butylated hydroxyanisole (BHA), butylated hydroxytoluene (BHT), and propyl gallate.
3. Spinach and rhubarb contain oxalic acid, which can cause kidney stones. Lima beans and sweet potatoes contain a substance that can be converted into toxic hydrogen cyanide (HCN) in the body.

Drill:

1. Which type of food additive does the following:
   a) provide smooth texture and consistency in the product
   b) prevent spoilage due to bacterial action
   c) prevent spoilage due to reactions with oxygen
2. Name a nutritional supplement commonly present in each of the following: a) milk, b) salt, c) flour

Objective 5: Explain some of the major issues concerning the safety of food additives.

- The Food and Drug Administration (FDA) is responsible in the United States for approving food additives. Many additives that have a long history of use without apparent harm are assumed to be safe until shown to be unsafe. New additives must be tested extensively for safety before being approved.
- One major concern with food additives is their potential to cause cancer or genetic damage.
- Judging whether a food additive is safe is sometimes difficult, as shown by the following questions:
  - To what extent do test results in animals apply to humans?
  - If an immense dose causes harm in animal tests, is this a reason to ban the use of very small amounts of the additive? How big a safety margin is desirable and practical?
  - To what extent should an additive be tested, not only by itself, but in combination with the other potential additives that might be present in a product?
  - To what extent should the metabolic products of additives be tested? Since animals may metabolize certain additives differently from humans, how reliable are animal tests?

Examples:

1. Some tests show that large doses of saccharin, an artificial sweetener, can cause cancer in test animals. Yet public pressure has caused saccharin to still be allowed as a food additive.
2. Sodium nitrite ($NaNO_2$) is used as a preservative in meats. There is some evidence that it is metabolized in the body to produce cancer-causing nitrosamines. Does its possible risk of causing cancer offset the protection it affords against botulism poisoning? This issue is not yet decided.
3. The FDA has banned several dyes from coal tar that were used as coloring agents. Those additives were found to cause cancer.

Drill:

1. The agency that approves food additives to be used in the United States is ________________________.
2. An artificial sweetener banned from use in foods and beverages is ____________.
3. Monosodium glutamate (MSG) is a controversial food additive in terms of safety. It is used in foods as a ____________ agent.

Self-Test

1. A food additive that disperses droplets of one liquid (such as oil) in another liquid (such as water) is a(n) a) acidulant, b) emulsifier, c) preservative, d) antioxidant, e) contaminant
2. A disease due to a shortage of calories and protein is a) night blindness, b) rickets, c) kwashiorkor, d) pellagra, e) anemia
3. An example of a fat-soluble vitamin is a) vitamin A, b) vitamin B_{12}, c) vitamin C, d) more than one of the above, e) none of the above
4. A controversial preservative in meats that protects against botulism poisoning is a) sodium nitrite, b) BHA, c) saccharin, d) MSG, e) none of the above
5. Two chemicals in combination may have a greater effect than the effects of each chemical used alone. This is called a ____________ effect.
6. An artificial sweetener that in some tests has been linked to cancer is a) sucrose, b) MSG, c) aspartame, d) sodium phosphate, e) saccharin
7. Proteins in the diet are digested to produce a) simple sugars, b) glycerol and fatty acids, c) DNA, d) amino acids, e) vitamins
8. Toxicity from too much of a vitamin is well known for a) vitamin D, b) vitamin C, c) vitamin E, d) vitamin B_{12}, e) none of the above
9. Most of the food additives that have been banned because of their potential to cause cancer are a) nutritional supplements, b) emulsifiers, c) coloring agents, d) alkalis, e) preservatives
10. True or False: A pure chemical substance isolated from a natural source is safer to consume than a pure sample of the same substance that was synthesized.
11. Goiter results from a deficiency of a) iodine, b) iron, c) calcium, d) sodium, e) phosphorus
12. A source of complete protein is a) corn, b) milk, c) peas, d) soybeans, e) potatoes
13. The major source of calories in the diets of most people is a) fats, b) carbohydrates, c) minerals, d) proteins, e) vitamins
14. A widely used antioxidant, particularly in fatty foods, is a) sodium nitrate, b) BHT, c) MSG, d) cyclamate, e) vitamin D
15. Slow blood clotting results from a deficiency of a) iodine, b) vitamin K, c) vitamin D, d) vitamin C, e) vitamin B_1

Answers

Objective 1:

1. c
2. b
3. a

Objective 2:

1. Protein that is adequate in amount may still be inadequate in one or more of the essential amino acids if the protein is not a complete protein.
2. b
3. c

Objective 3:

1. a
2. Cu (copper)
3. vitamin B_{12}, Fe (iron)

Objective 4:

1. a) stabilizer b) preservative c) antioxidant
2. a) vitamin D b) iodide ion (I^-) c) iron (Fe), various vitamins and amino acids

Objective 5:

1. Food and Drug Administration (FDA)
2. cyclamate
3. flavoring

Self-Test:

1) b; 2) c; 3) a; 4) a; 5) synergistic; 6) e; 7) d; 8) a; 9) c; 10) False; 11) a; 12) b; 13) b; 14) b; 15) b

Evaluation:

If you missed more than one question on the self-test in any of the following groups, you need to review the section indicated:

Question Groups	Section
2, 7, 12, 13	18.1
3, 8, 11, 15	18.2
1, 5, 10	18.3
4, 6, 9, 14	18.4

Chapter 19

Medical Drugs: Treating Diseases and Preventing Pregnancy

Outline

I. Major Types of Drugs: Fighting Disease
   A. Early Use of Chemicals to Treat Disease
   B. Types of Medical Drugs
      1. Chemotherapeutic drugs
      2. Metabolic drugs
      3. Nervous system drugs
II. Sulfa Drugs and Antibiotics: Fighting Infection
   A. Sulfa Drugs
   B. Penicillin and Other Antibiotics
      1. Discovery and action of penicillin
      2. Broad-spectrum antibiotics
      3. Problems with antibiotics
III. Antimetabolic Drugs: Treating Blood Clots, Gout, and Cancer
   A. Antimetabolites
   B. Treating Blood Clots and Gout
   C. Anticancer Drugs
      1. Antimetabolites
      2. Alkylating (cross-linking) agents
IV. Hormonal Therapy: Sending the Right Chemical Messages Throughout the Body
   A. Using Hormones to Treat Thyroid Disorders
   B. Other Examples of Hormone Therapy
      1. Adrenal hormones
      2. Prostaglandins
   C. Treating Diabetes
V. Steroids and the Pill: Preventing Pregnancy
   A. Steroid Hormones
      1. Estrogens
      2. Androgens
      3. Progestogens
   B. Birth Control in the Future

Objectives

After you read and study the chapter [and the sections in brackets], you should be able to:

[C] 1. Identify the three major types of medical drugs. [Section 19.1; Questions 1, 2]

[C] 2. Distinguish between sulfa drugs and antibiotics in terms of their origin, chemical structure, and mode of action as a drug. [Section 19.2; Questions 3, 4, 5, 6]

[C] 3. Identify what an antimetabolite is and how such compounds are used to inhibit blood clotting and to treat gout and cancer. [Section 19.3; Questions 7, 8, 9]

[C] 4. Explain how hormones or other chemicals are used to treat thyroid malfunction and diabetes. Explain how adrenal hormones and prostaglandins are used. [Section 19.4; Questions 10, 11, 12]

[C] 5. Explain how oral contraceptives were developed and how they work to prevent pregnancy. [Section 19.5; Questions 13, 14]

Practice

Objective 1: Identify the three major types of medical drugs.

Chemotherapeutic drugs, in the narrow sense of the term used in this chapter, are chemicals that kill or injure infectious organisms. "Chemotherapeutic drug" also may refer to any chemical used to treat disease or illness.
Metabolic drugs control, supplement, or substitute for various body chemistry processes.
Nervous system drugs affect the central nervous system or nerves that carry messages throughout the body.

Examples:

1. Penicillin is a chemotherapeutic drug because it treats certain types of infections.
2. An anesthetic such as Novocaine is a nervous system drug because it blocks the transmission of nerve impulses in a particular region of the body.
3. Tolbutamide, a metabolic drug, is used by some diabetics because it stimulates the pancreas to release more insulin; this helps lower their blood sugar levels.

Drill: Identify the following as a chemotherapeutic, metabolic, or nervous system drug:

1. tetracycline
2. Valium
3. cocaine
4. thyroid hormone pills

Objective 2: Distinguish between sulfa drugs and antibiotics in terms of their origin, chemical structure, and mode of action as a drug.

Sulfa drugs are synthetic substances, contain the structural group $(O{=}\underset{|}{\overset{|}{S}}{=}O)$, and kill microorganisms by blocking their metabolism of para-aminobenzoic acid (PABA) into folic acid.

Antibiotics typically are produced naturally by bacteria, molds, or fungi. They have a wide variety of chemical structures and kill microorganisms in a variety of ways; this includes interfering with protein synthesis or blocking the formation of cross-links in bacterial cell walls.

Examples:

1. 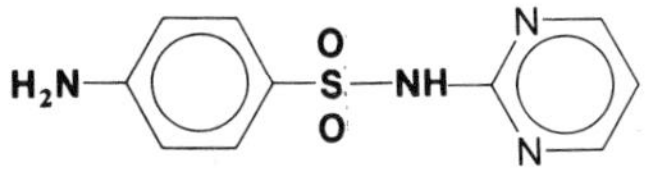 $H_2N-C_6H_4-S(O)(O)-NH-C_4H_3N_2$ (pyrimidine ring, N, N) is a sulfa drug.
2. Puromycin, a compound that blocks protein synthesis in certain bacteria, is an antibiotic.

Drill:

1. Are the complete structures of all antibiotics produced naturally?
2. What element(s) must sulfa drugs contain in their structures that antibiotics don't necessarily contain?
3. Suppose a new drug is discovered that kills certain bacteria by destroying their DNA. Would this more likely be a sulfa drug or an antibiotic?

Objective 3: Identify what an antimetabolite is and how such compounds are used to inhibit blood clotting and to treat gout and cancer.

Antimetabolites are compounds that structurally resemble a natural substance and thus interfere with the metabolism of that substance.

Antimetabolites that block the clotting action of vitamin K are anticoagulants. Antimetabolites that block the excessive production of uric acid from hypoxanthine are used to treat gout. Antimetabolites that block the making of new DNA thus block cell division; in some cases, they are used as anticancer drugs.

Examples:

1. Allopurinol, which structurally resembles hypoxanthine, is used to treat gout.
2. Warfarin, which structurally resembles vitamin K, is an anticoagulant.
3. Methotrexate is an anticancer drug that structurally resembles folic acid, a vitamin needed for the synthesis of new DNA when cells divide.

Drill:

1. How does 5-fluorouracil (Figure 19.15) work as an anticancer drug?
2. If too high a dose of an antimetabolite is taken, what can be used as an antidote?
3. If an antimetabolite were found that blocked the action of vitamin C (ascorbic acid), what disease might it cause? [Hint: See Section 18.2.]

Objective 4: Explain how hormones or other chemicals are used to treat thyroid malfunction and diabetes. Explain how adrenal hormones and prostaglandins work.

People with underactive thyroid glands (hypothyroidism) may be treated by administering thyroid hormone. If the low activity is due to a lack of iodine (a condition called goiter), adding iodide to the diet (as in iodized salt) is the usual treatment.

People with overactive thyroid glands (hyperthyroidism) may be treated with chemicals that inhibit the production of thyroid hormone.

Diabetes may be treated by insulin or by drugs that help stimulate the production of insulin.

Adrenal hormones work in various ways. Some relieve inflammations and rheumatoid arthritis; others affect salt and water retention in the body.

Prostaglandins are cyclic, fatty acid hormones that have a wide array of actions. They influence smooth muscle contraction, blood pressure, blood clotting, and allergic reactions.

Examples:

1. Thiocyanate ion (SCN^-) and perchlorate ion (ClO_4^-) block the incorporation of iodide ion (I^-) into thyroid hormone, and thus can be used to treat hyperthyroidism.
2. Tolbutamide, which stimulates the secretion of insulin, is used to treat diabetes.
3. Thromboxane, a prostaglandin, stimulates the formation of blood clots.

Drill:

1. What common adrenal hormones are used to treat inflammations and rheumatoid arthritis?
2. Diabetes is associated with low levels, or low activity, of the hormone ___________, which is normally secreted by the organ __________. Why can this hormone not be taken orally by diabetics?
3. A deficiency of thyroid hormone, produced by a nutritional lack of ___________, is called ___________.

Objective 5: Explain how oral contraceptives were developed and how they work to prevent pregnancy.

Oral contraceptives are synthetic forms of the female sex hormones, estrogens and progestogens. Two of the main chemical modifications in the natural hormones are 1) attaching an acetylene (-C≡CH) group to the five-carbon ring in the steroid structure and 2) removing a methyl group from the junction of the first two rings in the steroid structure of progestogens.

Oral contraceptives work by simulating a false pregnancy, thus preventing ovulation and conception.

Examples:

1. Norethindrone, a progesterone with the acetylene group attached and the methyl group removed, was one of the first substances used in birth control pills.
2. Mestranol, a synthetic estrogen, is used in some oral contraceptives.

Drill:

1. Why are synthetic, instead of natural, hormones used as contraceptives?
2. Why are estrogens and progestogens classified as steroids?
3. Which hormones--estrogens or progestogens--contain an aromatic ring? [Hint: See Figure 19.21.]

Self-Test

---

1. An oral hypoglycemic drug stimulates the body to produce more a) thyroid hormone, b) insulin, c) prostaglandins, d) cortisone, e) none of the above
2. An antimetabolite that is an anticoagulant works by blocking the clotting action of a) iodide, b) folic acid, c) uric acid, d) vitamin K, e) none of the above
3. To produce enough thyroid hormone, we need in our diets enough a) uric acid, b) iodide, c) folic acid, d) vitamin K, e) none of the above
4. A drug that kills bacteria by interfering with the synthesis of their cell walls is a) sulfanilamide, b) aminopterin, c) penicillin, d) Warfarin, e) none of the above
5. Some oral contraceptives contain a) tetracycline, b) a hypoglycemic drug, c) a synthetic estrogen, d) a prostaglandin, e) none of the above
6. Anticancer drugs often work by interfering with the functioning of a) DNA, b) para-aminobenzoic acid (PABA), c) bacterial cell walls, d) vitamin K, e) none of the above
7. Sulfa drugs work by structurally resembling a) folic acid, b) uric acid, c) PABA, d) cortisone, e) none of the above

8. Birth control pills are an example of a) chemotherapeutic drugs, b) metabolic drugs, c) nervous system drugs, d) antimetabolites, e) none of the above
9. Sulfa drugs are a) chemotherapeutic drugs, b) metabolic drugs, c) nervous system drugs, d) antimetabolites, e) none of the above
10. Fill in the blanks appropriately:
    - a) __________ hormones with a cyclic, fatty acid structure
    - b) __________ anticancer drug that resembles folic acid
    - c) __________ term for natural substance that kills or inhibits the growth of microorganisms
    - d) __________ substance produced in excess amounts in gout
    - e) __________ treatment for hypothyroidism
    - f) __________ example of a synthetic chemotherapeutic drug
    - g) __________ the two types of natural hormones that oral __________ contraceptives structurally resemble

---

Answers

Objective 1:

1. chemotherapeutic drug
2. nervous system drug
3. nervous system drug
4. metabolic drug

Objective 2:

1. No. Drug companies often make slight chemical changes in the naturally produced substance to make antibiotics that are even more effective.
2. sulfur and oxygen
3. Antibiotic. (Sulfa drugs block the formation of new DNA when bacteria reproduce.)

Objective 3:

1. By resembling a normal pyrimidine base in DNA, 5-fluorouracil blocks the synthesis of DNA. This stops the runaway, out-of-control cell division characteristic of cancer cells.
2. Giving a high dose of the natural substance that resembles the antimetabolite will serve as an antidote.
3. scurvy

Objective 4:

1. cortisone, hydrocortisone
2. insulin, pancreas
3. iodide, goiter

Objective 5:

1. Natural hormones cannot be taken orally because they are destroyed in the digestive tract.
2. They have the joined, four-ring structure characteristic of steroids.
3. estrogens

Self-Test:

1) b; 2) d; 3) b; 4) c; 5) c; 6) a; 7) c; 8) b; 9) a and (in bacteria) d; 10) a) prostaglandins, b) aminopterin, methotrexate, c) antibiotic, d) uric acid, e) thyroid hormone or iodide, f) sulfa drugs (such as sulfanilamide); g) estrogens, progestogens

Evaluation:

If you missed more than one question on the self-test in any of the following groups, you need to review the section indicated:

Question Groups	Section
8, 9, 10f	19.1
4, 7, 9, 10c, 10f	19.2
2, 6, 10b, 10d	19.3
1, 3, 10a, 10e	19.4
5, 8, 10g	19.5

Chapter 20

Chemistry and the Mind: Some Useful and Abused Drugs

Outline

I. Chemistry and the Nervous System: Sending the Right Messages
   A. What Is a Nerve Cell?
   B. How Are Nerve Impulses Transmitted?
II. Pain Killers and Depressants: Relieving Pain and Inducing Sleep
   A. Some Nervous System Drugs
   B. Mild Analgesics: Aspirin Is Aspirin Is Aspirin
   C. Strong Analgesics: Narcotics
      1. Narcotics
      2. Narcotic antagonists
      3. Endorphins and enkephalins
   D. Local Anesthetics
   E. General Anesthetics
   F. Sedatives
III. Mind Drugs: Treating Mental Diseases and Nerve Disorders
   A. Tranquilizers
   B. Chemistry and the Brain
      1. Treating Parkinson's disease
      2. Treating Alzheimer's disease
IV. Stimulants, Antidepressants, and Hallucinogens: Treating Depression and Tripping Out
   A. Stimulants and Antidepressants
      1. Drugs that inhibit monoamine oxidase
      2. Amphetamines
   B. Hallucinogens
      1. Natural compounds
      2. Synthetic compounds
      3. Marijuana
V. Drug Abuse: Getting Hooked and Unhooked
   A. Drug Abuse
      1. Psychological dependence
      2. Physical dependence
      3. Withdrawal
      4. Drug treatments for dependence
      5. Tolerance and cross-tolerance

Objectives

After you read and study the chapter [and the sections in brackets], you should be able to:

[C] 1. Explain the structure of nerve cells and how nerve impulses are transmitted in the body. [Section 20.1; Question 1]

[C] 2. Distinguish among mild analgesics, strong analgesics, local anesthetics, general anesthetics, and sedatives and know specific examples of each type of drug. [Section 20.2; Questions 2, 3, 4, 5, 6, 7, 8]

[C] 3. Explain what kinds of tranquilizers are used to treat manic depression and schizophrenia. Explain how drugs have been tried for treating Parkinson's disease and Alzheimer's disease. [Section 20.3; Questions 6, 9, 10]

[C] 4. Explain what types of drugs are used as stimulants, antidepressants, and hallucinogens and explain how they work. [Section 20.4; Questions 5, 6, 10, 11, 12]

[C] 5. Explain and distinguish among the following: types of drug dependence, withdrawal, tolerance, and cross-tolerance. Explain which drugs exhibit these characteristics. [Section 20.5; Question 6]

Practice

Objective 1: Explain the structure of nerve cells and how nerve impulses are transmitted in the body.

A neuron, or nerve cell, consists of a cell body, a long extension called an axon that sends nerve signals, and many branchlike extensions called dendrites that receive nerve signals.

To transmit a nerve impulse, a neuron releases a substance (called a neurotransmitter) from its axon that travels across a gap, or synapse, to bind to the dendrite of a neighboring neuron. The neurotransmitter causes Na^+ and K^+ ions to flow across the dendrite membrane, causing it to receive the nerve impulse from the axon of the nearby neuron. The neurotransmitter must then leave the dendrite so the neuron can be ready to receive another nerve impulse.

Examples:

1. Acetylcholine is a common neurotransmitter.
2. Noradrenaline, dopamine, and serotonin are other important neurotransmitters.

Drill:

1. The enzyme that removes acetylcholine from the dendrite of a neuron is called acetylcholinesterase. What would be the effect of a drug that blocked the action of this enzyme?
2. A nerve impulse travels from the __________ of a neuron across the __________ to a __________ of the receiving neuron.
3. Chemicals that trigger a flow of K^+ and Na^+ across the membrane of a neuron receiving a nerve impulse are called __________.

Objective 2: Distinguish among mild analgesics, strong analgesics, local anesthetics, general anesthetics, and sedatives and know specific examples of each type of drug.

Mild analgesics relieve moderate levels of pain.
Strong analgesics relieve intense pain at low levels. At high concentrations they typically produce stupor or sleep; such drugs are classified as narcotics.
Local anesthetics block pain in a specific area where they are applied.
General anesthetics produce unconsciousness and insensibility to pain.
Sedatives produce relaxation. If they produce sleep, they can be classified as hypnotics.

Examples:

1. Mild analgesics include aspirin, acetaminophen (Tylenol, Datril), and ibuprofen (Advil, Nuprin).
2. Strong analgesics include morphine, codeine, heroin, propoxyphene (Darvon), meperidine (Demerol), and methadone.
3. Local anesthetics include procaine (Novocaine), lidocaine (Xylocaine), dibucaine (Nupercaine), cocaine, and tetracaine (Pontocaine).
4. General anesthetics include nitrous oxide (N_2O), diethyl ether, halothane (Fluothane), enflurane (Ethrane), and sodium thiopental (sodium pentothal).
5. Sedatives and hypnotics include the barbiturates pentobarbital (Nembutal), secobarbital (Seconal), amobarbital (Amytal), and phenobarbital (Luminal).

Drill:

1. Which one of the five classes of compounds above is most widely used?
2. Classify the following according to the five types of drugs listed above: codeine, phenobarbital, laughing gas (N_2O), Demerol, Tylenol, and cocaine.
3. Which of the five classes of drugs above are the least likely to produce sleep?
4. Which of the five classes of drugs above have the most potential for physical dependence to develop?

Objective 3: Explain what kinds of tranquilizers are used to treat manic depression and schizophrenia. Explain how drugs have been tried for treating Parkinson's disease and Alzheimer's disease.

Tranquilizers relieve tension and anxiety by depressing the central nervous system; they also are used to treat people with schizophrenia or manic depression.

Parkinson's disease, a neural disorder involving shortages of the neurotransmitter dopamine in certain regions of the brain, is treated with dihydroxyphenylalanine (dopa), which is metabolized to dopamine.

Alzheimer's disease, a neural disorder more common in older people, is associated with shortages of acetylcholine in certain regions of the brain. Attempted treatments with substances that are metabolized to acetylcholine have not been successful.

Examples:

1. Widely used tranquilizers include chlorpromazine (Thorazine), promazine (Compazine), meprobamate (Equanil, Miltown), chlordiazepoxide (Librium), and diazepam (Valium).
2. Li^+ ion is used to treat hyperkinetic children, schizophrenia, and (especially) manic depression.
3. Choline, lecithin, and physostigmine have been tried (without success) in treating people with Alzheimer's disease.

Drill:

1. The only drug listed under the "Examples" above that is not an organic compound is ________.
2. Which group of drugs in the "Examples" above has the most potential for causing physical dependence?
3. Do tranquilizers increase or decrease the action of neurotransmitters in transmitting nerve impulses?

Objective 4: Explain what types of drugs are used as stimulants, antidepressants, and hallucinogens and explain how they work.

Antidepressants are drugs used to treat depression. Some are also classified as stimulants, and others (when used at higher concentrations) are tranquilizers. Many antidepressants work by increasing the levels of neurotransmitters in the brain; they do this by blocking the action of an enzyme, called monoamine oxidase (MAO), that normally breaks down neurotransmitters.

Stimulants are drugs that increase the activity of the brain and central nervous system and make people feel more alert and wakeful. Stimulants such as amphetamines work by increasing the action of certain neurotransmitters such as norepinephrine.

Hallucinogens are mind-affecting chemicals that cause vivid illusions, fantasies, and hallucinations. Since many hallucinogens structurally resemble serotonin, a neurotransmitter, they may act by altering nerve pathways that normally use serotonin.

Examples:

1. Antidepressants include isoniazid, imipramine (Tofranil), tranylcypromine (Parnate), and isocarboxazid (Marplan).
2. Stimulants include caffeine, amphetamine (Benzedrine), methamphetamine (Methedrine), and methylphenidate (Ritalin).
3. Hallucinogens include mescaline, psilocybin, LSD, marijuana, PCP, MDA, DOM (STP), and DMT.

Drill:

1. Which one of the following structurally resembles serotonin, a neurotransmitter: caffeine, Tofranil, Benzedrine, or mescaline.
2. Which one of the following inhibits the action of the enzyme, monoamine oxidase (MAO): caffeine, Tofranil, Benzedrine, or mescaline.
3. Which of the three classes of drugs above work by increasing the concentration or action of natural neurotransmitters?

Objective 5: Explain and distinguish among the following: types of dependence, withdrawal, tolerance, and cross-tolerance. Explain which drugs exhibit these characteristics.

Psychological dependence (habituation) is an intense desire to continue using a drug.

Physical dependence is a physical need to continue using a drug. When use of the drug is discontinued, the person experiences physical discomfort, called withdrawal, of various degrees of severity depending on the drug and the extent of use.

Tolerance occurs with increasing experience with a drug; higher doses of the drug are needed to produce the desired effect. Cross-tolerance may develop among related drugs, in which tolerance to one drug produces tolerance to other drugs that the person may not even have used.

Table 20.4 lists the potential of various drugs for physical dependence, psychological dependence, and tolerance. Drugs producing physical dependence typically produce withdrawal.

Examples:

1. Physical dependence occurs with the narcotics, ethanol, barbiturates, and caffeine.
2. Almost all of the drugs discussed in this chapter, except the mild analgesics, can cause tolerance to develop. All of these drugs have the potential to produce psychological dependence.
3. Cross-tolerance commonly occurs among the hallucinogens.

Drill:

1. For which of the following drugs is the evidence not clear that physical dependence develops: ethanol, caffeine, marijuana, Seconal, codeine.
2. A person who stops using a drug may experience physical discomfort for a while. Does this prove that the person had a physical dependence on the drug?
3. The development of tolerance can be dangerous. Why?

Self-Test

---

1. The most common drug used to treat Parkinson's disease is a) mescaline, b) dopa, c) meprobamate, d) isoniazid, e) none of the above
2. A common drug used to treat manic depression is a) dopa, b) psilocybin, c) chlorpromazine (Thorazine), d) Li^+ ion, e) none of the above
3. A drug that works by inhibiting monoamine oxidase (MAO) is a) ethanol, b) diazepam (Valium), c) phenobarbital (Luminal), d) acetaminophen, e) none of the above
4. An example of a strong analgesic is a) morphine, b) LSD, c) mescaline, d) aspirin, e) none of the above
5. Pentobarbital (Nembutal) is an example of a) mild analgesic, b) hypnotic, c) antidepressant, d) hallucinogen, e) strong analgesic
6. A drug that structurally resembles the neurotransmitter serotonin is a) psilocybin, b) caffeine, c) isoniazid, d) aspirin, e) none of the above

7. Cross-tolerance is likely to develop between psilocybin and a) secobarbital (Seconal), b) methamphetamine (Methedrine), c) heroin, d) mescaline, e) none of the above
8. A local anesthetic that is widely abused is a) LSD, b) marijuana, c) cocaine, d) ethanol, e) none of the above
9. N_2O, nitrous oxide, is used as a a) stimulant, b) anti-depressant, c) general anesthetic, d) mild analgesic, e) none of the above
10. Physical dependence can develop with the regular use of a) methadone, b) diazepam (Valium), c) secobarbital (Seconal), d) caffeine, e) none of the above
11. Which one of the following is <u>not</u> a neurotransmitter: a) serotonin, b) dopa, c) acetylcholine, d) noradrenaline
12. A drug that works as a mild analgesic by blocking the synthesis of prostaglandins is a) ibuprofen, b) mescaline, c) lidocaine (Xylocaine), d) phenobarbital (Luminal), e) none of the above
13. In order to receive a nerve impulse, a neuron needs to have a neurotransmitter substance bound to its a) dendrite, b) axon, c) cell body, d) nucleus, e) none of the above
14. A shortage of acetylcholine in certain regions of the brain is characteristic in people with a) schizophrenia, b) Parkinson's disease, c) hallucinations, d) Alzheimer's disease, e) none of the above
15. Psychological dependence may develop with the regular use of a) amphetamines, b) Cracker Jacks, c) PCP, d) codeine, e) none of the above
16. Neurotransmitters work by binding to the membranes of receiving cells, allowing the flow across those membranes of a) water, b) lithium ions, c) sodium and potassium ions, d) magnesium and calcium ions, e) none of the above
17. Withdrawal symptoms generally indicate a drug user has developed a) psychological dependence, b) physical dependence, c) tolerance, d) cross-tolerance, e) none of the above

---

Answers

<u>Objective 1:</u>

1. Such a drug would cause acetylcholine to remain on the dendrite, thus overstimulating the receiving neuron with nerve impulses.
2. axon, synapse, dendrite
3. neurotransmitters

<u>Objective 2:</u>

1. mild analgesics
2. strong analgesic, sedative/hypnotic, general anesthetic, strong analgesic, mild analgesic, local anesthetic

3. mild analgesics, local anesthetics
4. strong analgesics, sedatives/hypnotics

Objective 3:

1. Li^+ ion
2. the widely used tranquilizers (Example 1)
3. decrease

Objective 4:

1. mescaline
2. Tofranil
3. antidepressants and stimulants

Objective 5:

1. marijuana
2. Not necessarily. The psychological difficulty in not using the drug can produce physical symptoms, or the user may be very tired after discontinuing the use of stimulants. This is a major reason why it is difficult to determine whether drugs such as amphetamines, cocaine, and marijuana produce physical dependence.
3. As tolerance develops, the lethal dose may not change; therefore, the margin of safety for using the drug shrinks.

Self-Test:

1) b; 2) c and d; 3) e; 4) a; 5) b; 6) a; 7) d; 8) c; 9) c; 10) a-d; 11) b (dopa is metabolized to dopamine, which is a neurotransmitter); 12) a; 13) a; 14) d; 15) a-d; 16) c; 17) b

Evaluation:

If you missed more than one question on the self-test in any of the following groups, you need to review the section indicated:

Question Groups	Section
11, 13, 16	20.1
4, 5, 8, 9, 12	20.2
1, 2, 14	20.3
3, 6, 7	20.4
10, 15, 17	20.5

Chapter 21

Toxicology: Dealing With Poisons, Mutagens, Carcinogens, and Teratogens

---

Outline

I. Poisons and Toxicity Ratings: How Much Is Too Much?
  A. Definition of Poison and LD_{50}
  B. Types of Poisons
II. Corrosive and Metabolic Poisons: Destroying Skin, Enzymes, and the Blood's Oxygen Supply
  A. Corrosive Poisons
  B. Metabolic Poisoning from Cyanide
  C. Metabolic Poisoning from Carbon Monoxide
  D. Fluoride Poisoning
  E. Poisoning from Alcohols
III. Nerve Poisons: Causing the Ultimate Breakdown
  A. How Neurotoxins Disrupt Nerve Messages
  B. Examples of Neurotoxins
IV. Toxic Metals: Why the Mad Hatter and Sir Isaac Newton Went Mad
  A. Common Toxic Metals
  B. How Metals Are Toxic
  C. Chelation of Metals
V. Mutagens, Carcinogens, and Teratogens: Causing Genetic Errors, Cancers, and Birth Defects
  A. Mutations
  B. Mutagens
    1. What mutations are
    2. How different mutagens work
  C. Carcinogens
    1. Identifying carcinogens
    2. How cancer develops
  D. Teratogens

Objectives

After you read and study the chapter [and the sections in brackets], you should be able to:

[C] 1. Define "poison" and explain how poisons are rated in terms of their toxicity. [Section 21.1; Questions 1, 2, 3]

[C] 2. Distinguish between corrosive and metabolic poisons, give examples of each, and explain how they work. [Section 21.2; Questions 4, 5]

[C] 3. Explain how different types of neurotoxins work and give specific examples of each type. [Section 21.3; Question 6]

[C] 4. Identify the most common toxic metals and metal compounds, how they work as toxins, and how chelation therapy works. [Section 21.4; Questions 7, 8]

[C] 5. Distinguish among mutagens, carcinogens, and teratogens. Give examples of each type and explain how they work as toxic substances. [Section 21.5; Questions 9, 10, 11]

Practice

Objective 1: Define "poison" and explain how poisons are rated in terms of their toxicity.

A poison (toxic substance) is any chemical that causes illness or death in an organism in a relatively small amount.

Poisons are rated in terms of their LD_{50}, the dose (in mg per kg of body weight) that kills 50% of the subjects exposed to that dose.

Table 21.1 gives examples of poisons rated according to their LD_{50} values.

Examples:

1. Cyanide ion (CN^-) has an LD_{50} of 1 mg/kg body weight. According to Table 21.1, cyanide ion is rated as extremely toxic.
2. Ethylene glycol, the main component in permanent anti-freeze, has an LD_{50} of 8500 mg/kg body weight. According to Table 21.1, it is rated as slightly toxic.

Drill:

1. If it took a dose of 5 mg of a substance to kill half of a group of rats weighing 0.18 kg each, what would be the LD_{50} for this substance?
2. What would be the toxicity rating for the substance in Question 1 according to Table 21.1?
3. Are all chemicals poisons?

Objective 2: Distinguish between corrosive and metabolic poisons, give examples of each, and explain how they work.

Corrosive poisons damage or destroy tissues on contact. They may dehydrate tissues or destroy proteins, cell membranes, and other tissue materials.

Metabolic poisons interfere with a normal metabolic process. Often they inhibit an enzyme or other protein, or interfere with the action of DNA.

Examples:

1. Strong acids (such as sulfuric, hydrochloric, and nitric acid) and strong bases (such as sodium hydroxide and ammonia) are corrosive poisons.
2. Strong oxidizing agents such as hydrogen peroxide (H_2O_2) and sodium hypochlorite (NaOCl) are corrosive poisons.
3. Metabolic poisons include cyanide ion (CN^-), carbon monoxide (CO), fluoride ion (F^-), and fluoroacetate. In addition, various alcohols interfere with the metabolism of each other.

Drill:

1. Rotenone is a substance that kills certain types of insects by blocking their ability to use oxygen to produce energy. Is rotenone a corrosive or metabolic poison?
2. Recall (in Chapter 6) that a base will neutralize an acid. Since strong acids are corrosive poisons, would it be a good idea to put a strong base onto your skin to reduce the damage caused by a strong acid spilled on your skin?
3. Are corrosive poisons or metabolic poisons more likely to have a specific site of action and be treated by a specific antidote?
4. The metabolism of ethanol is inhibited by a) ethane, b) benzene, c) propanol, d) ammonia, e) sodium hypochlorite

Objective 3: Explain how different types of neurotoxins work and give specific examples of each type.

Neurotoxins disrupt the normal transmission of nerve signals. Three ways neurotoxins do this are to 1) lower the concentration of neurotransmitter substances by blocking their synthesis or accelerating their breakdown, 2) block the binding of neurotransmitters to receptor sites on dendrites, and 3) block the removal of neurotransmitters from dendrites once the nerve signal has been sent.

Examples:

1. Botulin toxin blocks the synthesis of acetylcholine, a neurotransmitter.
2. Local anesthetics (Novocaine, cocaine), curare, and atropine block the binding of acetylcholine to receptor sites on dendrites of nearby neurons.

3. Nerve gases, organophosphate and carbamate pesticides, and some mushroom toxins block the removal of acetylcholine from dendrites after the nerve signal has been transmitted.

Drill:

1. Which group of compounds in the three examples above overstimulate nerve pathways?
2. Which group of compounds in the three examples above prevent the transmission of nerve impulses?
3. Why does atropine (at low concentrations) work as an antidote for nerve gas?

Objective 4: Identify the most common toxic metals and metal compounds, how they work as toxins, and how chelation therapy works.

Common metals (and their compounds) that are toxic include antimony, arsenic (a metalloid), beryllium, cadmium, lead, manganese, mercury, and nickel.

Some metals are metabolic poisons; they interfere with metabolism by binding to -SH groups or elsewhere on enzymes, thus inactivating them. Toxic metals also may bind to other, nonenzyme proteins and inactivate them.

Chelating agents bind metal ions, thus preventing them from causing more damage.

Examples:

1. Mercury, arsenic, and lead are toxic because they inactivate enzymes and other proteins by binding to their -SH groups.
2. BAL (British Anti-Lewisite) is a chelating agent that effectively binds arsenic, mercury, lead, nickel, antimony, and manganese. It is used as an antidote for poisoning by these substances.

Drill:

1. Egg whites and milk contain proteins with -SH groups. How do these foods work as antidotes to metal poisoning?
2. Why is EDTA (ethylenediaminetetraacetic acid), a chelating agent, used instead of BAL to treat cadmium and beryllium poisoning?
3. Mercury is the most toxic in the form of a) metallic mercury, b) methyl mercury, c) mercury salts

Objective 5: Distinguish among mutagens, carcinogens, and teratogens. Give examples of each type and explain how they work as toxic substances.

Mutagens are substances that cause changes in the nucleotide sequence of DNA, called mutations. Changing the genetic information in this way causes the cell to produce faulty proteins.

Carcinogens cause <u>cancer</u>. Many are mutagens, but little else is known with certainty about how carcinogens cause cancer.

Teratogens are substances that cause birth defects. They interfere with normal embryonic and fetal development, especially during the first two to three months of pregnancy.

Examples:

1. Mutagens include polycyclic aromatic hydrocarbons, nitrous acid, and some nitrosamines and aldehydes (see Table 21.5).
2. Carcinogens include asbestos, vinyl chloride, dioxane, and some dyes and nitrosamines (see Table 21.6).
3. Teratogens include thalidomide, ethanol, methyl mercury, and possibly certain cadmium, arsenic, nickel, and chromium compounds.

Drill:

1. Can people with birth defects caused by teratogens pass those disorders on to their children?
2. Is a carcinogen also likely to be a mutagen or a teratogen?
3. Vinyl chloride, used in the production of polyvinyl chloride (PVC) is known to be a a) mutagen, b) carcinogen, or c) teratogen.

Self-Test

---

1. Organophosphate and carbamate pesticides are known to be a) carcinogens, b) neurotoxins, c) corrosive poisons, d) toxic metals, e) none of the above
2. A toxic material that typically inactivates proteins by binding to their -SH groups is a) atropine, b) nitric acid, c) mercury, d) thalidomide, e) none of the above
3. A polycyclic aromatic hydrocarbon, 3, 4-benzopyrene, is a a) carcinogen, b) neurotoxin, c) mutagen, d) teratogen, e) none of the above
4. The toxicity of a substance is typically expressed in terms of its a) concentration in the environment, b) residence time in the body, c) stability, d) LD_{50}, e) none of the above
5. A metabolic poison that prevents oxygen (O_2) from binding to hemoglobin is a) cyanide ion (CN^-), b) lead, c) carbon monoxide (CO), d) fluoride ion (F^-), e) none of the above
6. The single most valuable emergency treatment for an ingested poison is a) activated charcoal, b) milk, c) water, d) baking soda, e) raw eggs

7. Chelating agents such as BAL and EDTA are used to treat poisoning by a) nerve gas, b) thalidomide, c) methanol, d) strong acids, e) none of the above
8. A substance that is needed at low concentrations and toxic at higher concentrations is a) carbon monoxide (CO), b) fluoride ion (F^-), c) sulfuric acid (H_2SO_4), d) methyl mercury, e) none of the above
9. An example of a teratogen is a) hydrochloric acid (HCl), b) ethanol, c) botulin toxin, d) asbestos, e) none of the above
10. Fill in the blanks appropriately:
    a) ________ term for a substance that causes genetic changes in the cell
    b) ________ an antidote for nerve gas
    c) ________ a metalloid commonly included in a list of toxic "metals"
    d) ________ enzyme that breaks down acetylcholine after a nerve impulse has been transmitted
    e) ________ type of poison that damages or destroys tissues on contact
    f) ________ antidote for methanol poisoning
    f) ________ dose of a poison with an LD_{50} of 2 mg/kg body weight that would have a 50% chance of killing a person weighing 110 lb (50 kg)

Answers

<u>Objective 1:</u>

1. LD_{50} = 5 mg/0.18 kg = 28 mg/kg
2. very toxic (the LD_{50} is between 5 and 50 mg/kg)
3. No. Many chemicals in small amounts do not cause illness or death.

<u>Objective 2:</u>

1. metabolic poison
2. No. Strong bases are also corrosive poisons.
3. metabolic poisons
4. c

<u>Objective 3:</u>

1. compounds in Example 3
2. compounds in Examples 1 and 2
3. By blocking the binding of acetylcholine to dendrites, atropine stops the overstimulation of nerve pathways caused by nerve gases.

Objective 4:

1. The -SH groups in milk and egg whites bind to the toxic metal, keeping it from doing more damage.
2. EDTA binds cadmium and beryllium more tightly than BAL does.
3. b

Objective 5:

1. No. These disorders are not due to faulty DNA.
2. mutagen
3. carcinogen

Self-Test:

1) b; 2) c; 3) a,c; 4) d; 5) c; 6) a; 7) e; 8) b; 9) b;
10) a) mutagen, b) atropine, c) arsenic,
d) acetylcholinesterase (or cholinesterase), e) corrosive,
f) ethanol, g) 100 mg

Evaluation:

If you missed more than one question on the self-test in any of the following groups, you need to review the section indicated:

Question Groups	Section
4, 6, 10g	21.1
5, 8, 10e, 10f	21.2
1, 10b, 10d	21.3
2, 7, 10c	21.4
3, 9, 10a	21.5

Chapter 22

Better Bodies Through Chemistry: Tampering with Genes, Body Parts, and Athletic Performance

---

Outline

I. Genetic Engineering: Making New Organisms
   A. Splicing and Recombining DNA to Change the Genetic Instructions in a Cell
   B. Using Recombinant DNA Technology to Make Insulin
   C. Other Examples of Recombinant DNA Technology
II. Genetic Therapy: Curing Genetic Diseases
   A. Custom Designing of Humans: Can We Do It?
   B. Treating Genetic Diseases by Genetic Engineering
III. New Body Parts: Trading in Before It's Too Late
   A. The Partially Replaceable Person
   B. Artificial Joints, and Ear and Larynx Implants
   C. Plastic Surgery Implants, Blood Vessels, and Artificial Blood
   D. Artificial Kidneys and Livers
   E. Oxygenators and Insulin Pumps
   F. Artificial Hearts
IV. Better Athletic Performance: Increasing Oxygen Supply, Muscle, and Stamina
   A. More Oxygen and Less Muscle Soreness
      1. Oxygen debt
      2. Increasing the number of red blood cells
   B. Improving Body Fuel Supplies
      1. Carbohydrate
      2. Replenishing water
   C. Using Drugs to Improve Performance and Treat Injuries
      1. Anabolic steroids
      2. Caffeine
      3. DMSO

Objectives

After you read and study the chapter [and the sections in brackets], you should be able to:

[C] 1. Explain how desired genes can be inserted into DNA and be used to produce needed protein materials such as insulin and vaccines. [Section 22.1; Questions 1, 2]

[C] 2. Explain how genetic engineering could be used to treat people with genetic disorders. [Sections 22.1, 22.2; Questions 4, 5]

[C] 3. Identify artificial body parts in common use, and discuss the advantages and limitations of artificial parts. [Section 22.3; Questions 5, 6, 7]

[C] 4. Explain how athletic performance can be improved by increasing oxygen capacity, maintaining optimal carbohydrate and water levels, and the use of drugs. [Section 22.4; Questions 8, 9, 10, 11]

Practice

Objective 1: Explain how desired genes can be inserted into DNA and be used to produce needed protein materials such as insulin and vaccines.

A gene is a section of DNA that carries the information for making one protein.

A plasmid is a circular piece of DNA; bacterial cells typically contain plasmids. A virus consists of DNA or RNA surrounded by a protein coat. Plasmids and viruses (of the DNA type) are the main vectors (vehicles) to carry designed genes into cells.

Recombinant DNA is DNA containing DNA from two or more different sources. It is typically made by treating a vector (plasmid or virus) with a restriction enzyme, which cuts the DNA at a very specific place, leaving the DNA pieces with "sticky ends." The desired gene is treated in the same way, giving it the same "sticky ends." When this treated gene is put together with the treated vector, they join together to form recombinant DNA containing the desired gene. This recombinant DNA then enters the target cell, carrying in the desired gene(s). Figure 22.3 shows how this works.

This technology has been used to make microorganisms into miniature factories to produce desired proteins such as insulin. Microbes can also be designed to produce proteins that stimulate the immune system to make antibodies against bacteria or viruses that contain those proteins; such proteins work as vaccines.

Examples:

1. Plasmids are the usual vectors for bacterial and yeast cells; viruses are more common in supplying animal cells with new genes.
2. Hemoglobin contains two different protein chains, so two different genes are needed to code for its proteins.
3. Recombinant DNA has been used to make human growth hormone and a vaccine against one form of hepatitis.

Drill:

1. The usual vectors for supplying new genes to cells are ________ and ________.
2. Given the appropriate genes, bacterial cells are capable of producing which of the following types of proteins: a) animal, b) plant, c) bacterial, d) all of the above, e) none of the above
3. Proteins that stimulate the immune system to make antibodies can be used as a) hormones, b) enzymes, c) vaccines, d) none of the above
4. DNA containing material from two or more different sources is called __________.

Objective 2: Explain how genetic engineering could be used to treat people with genetic disorders.

People with a genetic disorder due to the lack of one gene are the most likely candidates for gene therapy. People with genetic disorders involving multiple genes, or extra genetic material, are not likely candidates.

Gene therapy hasn't yet been done successfully in a person. It may become possible to supply people with a needed gene in the form of recombinant DNA, using a virus as the vector. The problems will include getting an adequate dose to the right cells and having the gene(s) function appropriately.

Examples:

1. Cystic fibrosis, phenylketonuria (PKU), albinism, and sickle cell anemia are genetic disorders due to a single abnormal gene. These are potential candidates for gene therapy.
2. Down's syndrome is due to extra chromosomal material and is not currently a candidate for gene therapy by recombinant DNA. Neither are spina bifida and cleft palate, which may involve the action of several genes plus environmental factors.

Drill:

1. Would birth defects due to teratogens (Chapter 21) such as thalidomide and ethanol be potential candidates for gene therapy? Why?
2. Hemophilia A is a disorder in which the blood doesn't clot properly. It occurs mostly in males and is caused by the lack of one gene. Might it be a potential candidate for gene therapy?
3. Would a person with a genetic disorder who was treated successfully with gene therapy still be able to transmit that disorder to her or his children that were born after the gene therapy?

Objective 3: Identify artificial body parts in common use, and discuss the advantages and limitations of artificial parts.

Artificial joints include hips, wrists, fingers, elbows, shoulders, and knees. Ear and larynx implants, artificial blood and blood vessels, and silicone implants for cosmetic purposes also are widely available.

Devices to replace all, or part, of an organ's functions include "artificial kidney" dialysis units, respirators, membrane oxygenators, insulin pumps, and artificial hearts and heart-assist devices.

In comparison with natural body parts, artificial parts often are less vulnerable to infection and degenerative diseases; they also may be less versatile, incompatible with body tissues, not adaptable to growth, and incapable of self-repair.

Drill:

1. An insulin pump supplements the normal action of which organ?
2. What technical factors make an artificial liver a less likely prospect than an artificial heart?
3. What are some practical uses for artificial blood?
4. Can artificial parts use the body's normal source of chemical energy?

Objective 4: Explain how athletic performance can be improved by increasing oxygen capacity, maintaining optimal carbohydrate and water levels, and the use of drugs.

When intense muscular activity causes a shortage of oxygen to oxidize carbohydrates, fats and proteins, less energy is produced in the body; this condition is called oxygen debt.

Increased oxygen capacities can be developed by physical exercise (to expand heart and lung capacity), and by increasing the number of red blood cells by exposure to high elevations or by receiving a dose of one's own blood cells, a practice commonly called blood doping.

Maintaining high carbohydrate and water levels is important for extended physical exercise. Glycogen levels can be increased by depleting glycogen through exercise or diet and then eating carbohydrate-rich foods for about 3 days prior to the exercise; this is called carbohydrate (or glycogen) loading. Water and carbohydrate levels are maintained during exercise by ingesting appropriate food and drinks.

Many different drugs are used (sometimes illegally) to improve athletic performance. They include drugs to increase muscle mass, treat injuries, and serve as stimulants.

Examples:

1. Athletic drinks such as Gatorade and E.R.G. contain water, glucose, and salts to replenish the loss of these materials during strenuous exercise.
2. Before a marathon, many runners do carbohydrate loading. They also may drink some coffee or other caffeine source an hour or so before the run.
3. Some weight lifters and body builders take anabolic steroids to increase their muscle mass. These drugs are dangerous and illegal for competition.

Drill:

1. The use of anabolic steroids is associated with an increased risk of a) diabetes, b) liver cancer, c) heart disease, d) kidney disease, e) epilepsy
2. Oxygen debt is associated with a) aerobic exercises, b) increased efficiency of energy production, c) poor nutrition, d) production of lactic acid, e) reduced energy production
3. Oxygen capacity can be increased substantially by a) eating more natural foods, b) blood doping, c) caffeine, d) living at sea level, e) all of the above
4. In addition to water and salts, athletic drinks such as Gatorade commonly supply a) sugar, b) vitamins, c) protein, d) fats, e) all of the above

---

Self-Test

---

1. A dialysis machine that removes impurities from the blood is helping take over the normal functions of the a) heart, b) lungs, c) kidneys, d) pancreas, e) none of the above
2. There is some evidence that endurance is increased by the use of a) DMSO, b) caffeine, c) ethanol, d) ethyl chloride, e) protein
3. A genetic disorder characterized by the presence of extra chromosomal material is a) phenylketonuria (PKU), b) sickle cell anemia, c) muscular dystrophy, d) hemophilia, e) none of the above
4. A circular piece of DNA often used as a vector in recombinant DNA experiments is a a) plasmid, b) restriction enzyme, c) virus, d) palindrome, e) none of the above
5. A membrane oxygenator temporarily substitutes for the normal action of the a) heart, b) lungs, c) kidneys, d) pancreas, e) none of the above

6. Gene therapy is most feasible to a) remove a faulty gene, b) reduce the number of chromosomes, c) supply a needed gene, d) supply a needed chromosome, e) all of the above
7. Recombinant DNA already is being used to produce a) steroid hormones, b) penicillin, c) DMSO, d) human growth hormone, e) none of the above
8. Muscle mass is increased by a) testosterone, b) caffeine, c) amphetamines, d) carbohydrate loading, e) blood doping
9. Fill in the blanks appropriately:
   a) ________ term for DNA containing DNA from two or more different sources
   b) ________ major nutrient source of energy (Calories) during very intense physical activity
   c) ________ type of genetic disorder that must be inherited from both parents as carriers
   d) ________ term for injecting a supply of one's own red blood cells shortly before competition
   e) ________ body part for which an artificial joint is most widely used
   f) ________ substance that opens plasmids at specific places, leaving them with "sticky ends"
   g) ________ material used as artificial blood

Answers

<u>Objective 1:</u>

1. plasmids, viruses
2. d
3. c
4. recombinant DNA

<u>Objective 2:</u>

1. No. These disorders are not genetic.
2. Yes
3. Yes, unless gene therapy also modified the reproductive cells (ova or sperm-producing cells).

<u>Objective 3:</u>

1. pancreas
2. The liver carries out so many diverse metabolic processes that no artificial device is likely to be able to do all of this. In contrast, the function of a heart is relatively simple; it is primarily a pump.
3. Artificial blood is used in emergencies when blood, or the needed blood type, isn't available. It can also be used for people who oppose blood transfusions for religious reasons.

4. No. Artificial parts that need an energy supply must have an external (to the body) source.

Objective 4:

1. b and c
2. d and e
3. b
4. a

Self-Test:

1) c; 2) b; 3) e; 4) a; 5) b; 6) c; 7) d; 8) a;
9) a) recombinant DNA, b) carbohydrate, c) recessive,
d) blood doping, e) hip, f) restriction enzyme,
g) perfluorodecalin (Fluosol)

Evaluation:

If you missed more than one question on the self-test in any of the following groups, you need to review the section indicated:

Question Groups	Section
4, 7, 9a, 9f	22.1
3, 6, 9c	22.2
1, 5, 9e, 9g	22.3
2, 8, 9b, 9d	22.4

GLOSSARY

Acid rain Droplets of nitric acid and sulfuric acid solutions that form when water reacts with common air pollutants such as nitrogen and sulfur oxides; acid deposition.

Acid-base reactions Reactions in which a proton (H^+) is transferred from an acid or proton donor to a base or proton acceptor.

Acidic solutions Water solutions in which there are more hydronium (H_3O^+) ions than hydroxide (OH^-) ions.

Acids Any hydrogen-containing substances that when dissolved in water donate hydrogen (H^+) ions or protons to water or another substance.

Active solar heating Capturing and storing the sun's energy in any type of dedicated solar collector.

Addiction A strong dependence on a chemical that produces pronounced physical or emotional reactions when withdrawn.

Addition polymerization Polymerization in which monomers combine or link together without the loss of any atoms.

Aerobic organisms Organisms that require oxygen to live.

Afterburners Pollution control devices for automobiles that recirculate exhaust gas into high temperature chambers to convert carbon monoxide to carbon dioxide and to burn any remaining hydrocarbons.

Air pollution Air that has chemicals or heat in high enough concentrations to harm humans, other animals, vegetation, or materials.

Alcohols Organic compounds with at least one hydroxyl (-O-H) group attached to a carbon atom (but not to a carbonyl group), R-O-H.

Aldehydes Compounds containing a carbonyl group attached to at least one hydrogen: $\mathrm{R\text{-}\underset{\underset{O}{\|}}{C}\text{-}H}$.

Alkali metals Elements (except hydrogen) located in column 1 or IA on the periodic table of the elements.

Alkaline earths Elements located in column 2 or IIA on the periodic table of the elements.

Alkaline solutions Water solutions in which there are more hydroxide (OH^-) ions than hydronium (H_3O^+) ions; a basic solution.

Alkanes Saturated hydrocarbons; contain carbon-carbon single bonds only.

Alkenes Hydrocarbons with one or more carbon-carbon double bonds.

Alkylating agents Anticancer drugs that bind to DNA and prevent it from functioning normally.

Alkynes Hydrocarbons with one or more carbon-carbon triple bonds.

Alpha particle A package of two protons and two neutrons emitted by a radioactive substance; the nucleus of a helium-4 atom with a 2+ charge.

Alzheimer's disease A disease characterized by loss of short-term memory, confusion, and eventually, loss of ability to read, write, calculate, and speak.

Amides Organic compounds containing nitrogen attached to a carbonyl group: **R**-N(**R'**)-C(=O)-

Amines Organic compounds containing nitrogen that is not attached to a carbonyl group: **R**-N(**R"**)-**R'**.

Amino acids Compounds containing a central carbon atom attached to a hydrogen, a carboxylic acid group, and amino group, and a fourth, miscellaneous, **R**-group: H-N(H)-CH(**R**)-C(=O)-O-H.

Amphetamines Stimulants with a characteristic amine structure (Figure 20.21).

Anabolic steroids A group of steroids, which includes testosterone and other related compounds (Figure 22.16), that stimulate protein synthesis and are used illegally to increase muscle mass.

Anaerobic organisms Organisms that do not require oxygen to live.

Analgesia Pain relief; one possible effect of a nervous system drug.

Androgens Male sex hormones secreted by the testes.

Anionic surfactants Surfactants that have a negative charge on their water-soluble heads.

Anode The positive electrode.

Antibiotics Chemicals produced by organisms that kill or inhibit the growth of infectious microorganisms.

Anticoagulants Drugs that prevent the normal clotting of blood in an animal.

Antidepressants Drugs that relieve the feeling of depression.

Antimetabolites Drugs that interfere with some normal biological process in an organism.

Antiperspirant A product that prevents body odor by inhibiting perspiration or preventing it from reaching the skin.

Aromatic hydrocarbons Certain cyclic or ring compounds with one or more double bonds; especially those containing the six-membered benzene ring (Figure 8.3).

Atmosphere The gaseous envelope of air surrounding the earth.

Atom The smallest particle that retains the chemical properties of an element.

Atomic mass The average mass in atomic mass units of all the isotopes found in a naturally occurring sample of an element, weighted to reflect the abundance of each isotope; atomic weight.

Atomic mass unit (amu) A mass exactly equal to one-twelfth the mass of a carbon-12 atom.

Atomic number The number of protons in the nucleus of an atom; the identifying characteristic of an element.

Atomic theory The fundamental concept that the basic structural unit of all matter is the atom.

Avogadro's number The number of structural particles in one mole of an element or compound; 6.02×10^{23}.

Axon The long fiber transmitter at one end of a nerve cell.

Background radiation That amount of radiation that is always and unavoidably present from natural sources.

Barbituates Sedatives and hypnotics derived from barbituric acid.

Bases Substances that when dissolved in water accept hydrogen (H^+) ions or protons from another substance.

Basic solutions Water solutions in which there are more hydroxide (OH^-) ions than hydronium (H_3O^+) ions; an alkaline solution.

Battery A source of electrical current created by an oxidation-reduction reaction that converts chemical potential energy to electrical energy.

Beta particle An electron emitted by a radioactive substance.

Biochemical oxygen demand (BOD) The amount of dissolved oxygen needed to consume organic wastes in water.

Biochemistry Life chemistry; the branch of chemistry that studies the composition, structure, and reactions of the materials of living organisms.

Biodegradable Able to be degraded by bacteria in the environment; said of a substance, especially a surfactant.

Biofuels Gaseous or liquid fuels that have been synthesized from organic matter or biomass.

Biomass Organic matter that can be burned directly as fuel or converted to more convenient gaseous or liquid biofuels.

Biopolymers Polymers that take part in biological processes.

Biosphere All forms of plant and animal life on earth.

Bleaches Oxidizing agents that render stains colorless.

Blood doping The practice of receiving injections of one's own red blood cells from previously donated blood to boost the oxygen-carrying capacity of the blood for competition.

Bohr model of the atom The atom is a nucleus surrounded by electrons that reside in certain allowed energy levels only.

Boiling point The temperature at which a liquid boils into a gas.

Boiling The physical change of state from liquid to gas; the opposite of condensation.

Breeder reactors Reactors that produce energy by fission and at the same time make new fissionable fuel in the form of plutonium-239 from nonfissionable but abundant uranium-238.

Broad-spectrum antibiotics Antibiotics effective against a wide variety of bacteria.

Builders Water-conditioning ingredients in a synthetic laundry detergent.

Calorie A unit of energy, especially food energy.

Cancer Collective name for a group of more than 100 different diseases characterized by uncontrolled cell growth and reproduction.

Carbamate pesticides Chemicals derived from carbamic acid, the molecule made by attaching an amino group to the carbon of a carboxylic acid group.

Carbohydrate loading The practice of depleting one's glycogen levels (typically by heavy exercise) then boosting the levels above normal just before competition by changing to a carbohydrate-rich diet.

Carbohydrates Cyclic compounds of carbon, hydrogen, and oxygen that contain several alcohol or hydroxyl (-OH) groups.

Carbon cycle The transfer of carbon between living and nonliving things in the environment (Figure 10.3).

Carbonyl group The group consisting of an oxygen atom double bonded to a carbon atom.

Carboxylic acid Compounds containing a carbonyl group attached to an oxygen atom that in turn is attached to a hydrogen atom: $R-\underset{\substack{\| \\ O}}{C}-O-H$; an organic acid.

Carcinogens Chemicals that cause cancer.

Catalysts Substances that increase the rate of a reaction without being consumed in the reaction.

Catalytic converters Pollution control devices for automobiles that mix exhaust gases with outside air in a chamber cotaining catalysts to ensure complete combustion.

Cathode ray tube Gas discharge tube; a device that creates cathode rays or electrons.

Cathode rays Streams of electrons.

Cathode The negative electrode.

Cationic surfactants Surfactants that have a positive charge on their water-soluble heads.

Cellulose An indigestible polymer of glucose that is the main structural material of plants and trees.

Chain reaction The runaway nuclear process by which any one reaction produces reactants for several others.

Chelating agents Chemicals that can bind to a metal with two bonds in a clawlike fashion.

Chemical bonds The forces holding atoms together in a molecule.

Chemical change Any change that alters the chemical composition of a substance changing it to a different substance.

Chemical energy The potential energy of the chemical bonds between atoms.

Chemical equation The shorthand representation of a chemical reaction which identifies all reactants and products and which indicates that mass is conserved.

Chemical reaction A chemical change in which one or more reactants are changed into one or more products.

Chemistry The scientific study of matter, its properties, and its reactions especially when using the atomic theory as the fundamental model.

Chemotherapeutic drugs Drugs that kill or injure infectious organisms.

Chemotherapy The use of chemicals to treat disease, especially diseases caused by microorganisms.

Chlorinated hydrocarbons Hydrocarbons with one or more chlorine atoms attached to a carbon atom.

Cholesterol A particular steroid that is used by the body to make vitamin D and sex hormones (Figure 9.11).

Coal A black, combustible, mineral solid containing 55 to 95% carbon.

Coal gasification The process of converting coal into a cleaner-burning gas that can travel through pipelines.

Coal liquefaction The process of converting coal into liquid fuels such as methanol or synthetic gasoline.

Collision energy The combined energy of molecules that run into each other.

Collision frequency The number of collisions between molecules during a given period of time.

Collision orientation The relative position or geometry between two molecules at the instant of their collision.

Commercial inorganic fertilizers Plant nutrients obtained from rock deposits of nitrates, phosphates, and other minerals, or synthesized from their elements.

Complex sugars Carbohydrates that contain more than one ring; polymers of the simple sugars; polysaccharides.

Compounds Pure substances composed of two or more elements combined chemically in fixed proportions.

Condensation polymerization Polymerization in which one or more small substances (such as water) are split out as monomers link together to form polymers.

Condensation The physical change of state from gas to liquid; the opposite of boiling.

Contact herbicides Chemicals that kill plants on contact by interfering with their photosynthesis.

Continuous spectrum The separation of white light into all the colors of the rainbow.

Control rods Parts of a nuclear fission reactor that capture neutrons and allow operators to regulate the nuclear reaction.

Copolymers Polymers made up of two or more different monomer units.

Corrosive poisons Poisons, such as strong acids, bases, and oxidizing agents, that destroy tissue on contact.

Covalent bonds The attraction between two atoms for a share of an electron pair.

Critical mass The amount of fissionable material necessary to sustain a nuclear chain reaction.

Cross-tolerance The development of a tolerance for related drugs without taking them.

Crude oil Petroleum; a dark, greenish-brown, foul-smelling liquid containing mostly hydrocarbons.

Decay curve A plot of the amount of a given radioisotope remaining after any specific time.

Demineralization The dissolving of calcium, phosphate, and hydroxide ions from tooth enamel in saliva; tooth decay; the opposite of mineralization.

Denatured alcohol Ethyl alcohol made unfit to drink by the addition of small amounts of toxic or foul-smelling compounds.

Dendrite One of the many antennalike receivers attached to a nerve cell.

Deodorant A product that prevents body odor by killing odor-causing bacteria, decomposing the foul-smelling compounds, or masking the odor with more pleasant fragrances.

Desalination Any process that removes salt from water.

Detergent action The ability of a surfactant to loosen grease from fabrics, dishes, or skin and to suspend it in rinse water.

Digestion The chemical splitting of food or nutrient molecules into the pool of simpler compounds that the body uses.

Dipole A molecule with an unequal distribution of charge so that one side of the molecule has a partial negative charge and the other has a partial positive charge.

Dipole-dipole interactions The electrostatic forces of attraction between the oppositely charged ends of two dipoles.

Disaccharides Carbohydrates that contain two rings.

Dissolved oxygen The amount of oxygen dissolved in water for use by aquatic plants and animals.

Distillation The process by which a solution is boiled and then condensed to produce pure solvent.

DNA Deoxyribonucleic acid; the polymer of nucleotides that contains all the genetic information of your body.

Drug abuse The ingestion of any chemical in a way that is not medically or legally approved in a given culture.

Drug tolerance A drug user's need to take larger and larger doses to achieve the same effects.

Dynamic equilibrium The condition in which the rates of two opposing processes are equal, and, while both reactions continue, no net change occurs.

Elastomers Polymers, like rubber, that have long, flexible chains that can be coiled and uncoiled.

Electrodes Conductors of electricity.

Electron dot structure The chemical symbol of an element with one dot for each of its valence electrons written next it.

Electron The subatomic particle found outside the nucleus that carries the negative charge in an atom.

Electronegativity The tendency of an atom to be the negative side of a bond; the tendency of an atom to attract an electron pair.

Electronic structure The particular arrangement of electrons in the energy levels of an atom.

Electrostatic forces The forces of attraction or repulsion between electrically charged bodies.

Elements Pure substances that cannot be broken down into simpler substances by ordinary chemical or physical means; pure substances that are found on the list of elements; collections of atoms with the same atomic numbers.

Emollient An ingredient of moisturizers that prevents water from evaporating out of the skin.

Endothermic reaction A reaction that absorbs or requires energy from the surroundings.

Energy conservation The saving of high-quality energy for non-wasteful uses.

Energy efficiency The percentage of total energy put into a task that does useful work and is not converted to waste heat.

Energy The capacity to move matter or to cause it to undergo change.

Entropy A measure of disorder or randomness.

Enzymes Protein molecules that act as catalysts for specific biological reactions.

Equilibrium The condition in which the rates of two opposing processes are equal, and, while both reactions continue, no net change occurs.

Essential amino acids Those eight amino acids that the human body cannot synthesize and must be supplied ready-made in the diet (Table 18.2).

Esters Compounds containing a carbonyl group attached to an oxygen atom that in turn is attached to a carbon atom: $R{-}\underset{\underset{O}{\|}}{C}{-}O{-}R'$.

Estrogens Female sex hormones produced mainly in the ovaries.

Ethers Organic compounds with at least one oxygen attached between two carbon atoms; $R{-}O{-}R'$.

Eutrophication The natural process by which lakes gradually become enriched in plant nutrients, grow blue-green algae, lose dissolved oxygen, and fill with plant sediment.

Excited state An electronic structure of an atom that has energy greater than that of the ground state and is therefore unfavorable and temporary.

Exothermic reaction A reaction that gives off energy to the surroundings.

Family Each of the eighteen vertical columns in the periodic table of the elements; a group.

Fatty acids Carboxylic acids with an even number of carbon atoms, varying from 12 to 24; components of triglycerides.

Fermentation The process by which microorganisms convert sugars into ethyl alcohol in the absence of air.

Fertilizers Substances added to soil to correct deficiencies in plant nutrients.

First law of energy or thermodynamics Energy is neither created nor destroyed in any process but merely transformed from one form to another.

Flame tests The characteristic color emitted by elements when samples of them are heated in flames.

Fluorosis Dark stains and pitting of the teeth caused by ingestion of excessive amounts of fluoride.

Flux A chemical that reacts with nonmetallic impurities in an ore to produce slag.

Foaming The formation of suds or bubbles in a liquid when there is more surfactant than will fit on its surface.

Formula unit The basic repeating unit in an ionic compound.

Fossil fuels The source of naturally occurring hydrocarbons: natural gas, coal, and petroleum.

Fractional distillation The process that uses the differences in boiling points among compounds to separate a mixture of them into pure substances.

Free radicals Molecules with an odd number of electrons which are often very reactive.

Freezing point The temperature at which a liquid freezes into a solid; the same temperature as the solid's melting point.

Freezing The physical change of state from liquid to solid; the opposite of melting.

Fuel cell A source of electrical current created by an oxidation-reduction reaction to which reactants are supplied continuously.

Fuel rods Parts of a nuclear fission reactor that contain the fissionable material consumed as fuel.

Functional groups Reactive groups of atoms attached to an organic molecule that determine most of the resulting product's chemical behavior.

Gamma rays High energy electromagnetic radiation emitted by a radioactive substance.

Gas The state of matter characterized by arbitrary volume and arbitrary shape.

Gasoline That fraction of the crude oil mixture that is used as a motor fuel.

General anesthesia Complete insensibility to pain and reversible unconsciousness; one possible effect of a nervous system drug.

Genes Segments of DNA within a chromosomes of a cell; they carry the information necessary for making the specific proteins an organism needs.

Genetic engineering Technology to change the genetic instructions in a cell so that it will make proteins that are different from the ones it ordinarily makes.

Geothermal energy Heat produced by lava in the interior of the earth that escapes through hot springs, geysers, volcanoes, and other natural phenomena.

Glucose The monosaccharide that serves as the fuel for your body (Figure 9.3).

Glycogen A polymer of glucose found in the liver. It is one form of energy storage in your body.

Greenhouse effect The heat-holding effect of gases in the atmosphere, such as CO_2, that allow infrared radiation to strike the earth but do not let it radiate back into space.

Ground state The electronic structure of an atom that has the least amount of energy and is therefore the most favorable.

Groundwater Liquid water that exists underground.

Group Each of the eighteen vertical columns in the periodic table of the elements; a chemical family.

Half-life The amount of time it takes for one half of the nuclei of a specific radioisotope to decay.

Hallucinogens Mind-affecting chemicals that cause vivid illusions, fantasies, and hallucinations.

Halogens Elements (except hydrogen) located in column 17 or VIIA on the periodic table of the elements.

Hard water Water that contains dissolved mineral ions such as Ca^{2+}, Mg^{2+}, or Fe^{2+}.

Hazardous waste Any discarded material that may pose a substantial threat to human health or to the environment when handled improperly.

Heat capacity A measure of how much energy is required to raise the temperature of a given mass of a substance.

Heat of vaporization The amount of energy required to vaporize a given quantity of a substance.

Herbicides Pesticides targeted against weeds.

Heterogeneous matter Matter having different parts with different properties and composition.

Homogeneous matter Matter with the same properties and composition throughout.

Homogeneous mixture Two or more pure substances blended together such that all parts of the combination have the same properties and composition; a solution.

Hormones Chemicals produced by cells in organisms that travel in the bloodstream and control various aspects of their growth and development.

Humectant An ingredient of moisturizers that attracts water to the skin from the outer surroundings.

Hydrocarbons Organic compounds of carbon and hydrogen only.

Hydrogen bonding The force of attraction between a hydrogen atom bonded to a fluorine, oxygen, or nitrogen atom with an F, O, or N atom elsewhere.

Hydrosphere The total quantity of water on earth, in the liquid, solid, and gaseous states.

Hyperglycemia The abnormal condition of having too much glucose in the bloodstream; often caused by diabetes.

Hyperthyroidism The condition caused by an overactive thyroid gland which makes its victim slender and active.

Hypnotic action Calming with sleep; one possible effect of a nervous system drug.

Hypoglycemia The abnormal condition of having too little glucose in the bloodstream.

Hypoglycemic drugs Drugs that lower a person's blood sugar levels.

Hypothyroidism The condition caused by low thyroid activity which makes its victim obese and sluggish.

Incomplete proteins Any proteins (especially plant protein) that lack sufficient quantities of one or more of the essential amino acids.

Industrial smog Air pollution from sulfur dioxide and particulate matter.

Inhibitors Substances that decrease the rate of a reaction without being consumed in the reaction.

Initiator One of the two categories of carcinogens that must act jointly to cause cancer; needed only in a brief, single dose, after which a promoter can develop the cancer.

Insecticides Pesticides targeted against insects.

Intermolecular forces The forces of attraction between molecules as opposed to chemical bonds which are the forces of attraction between atoms within molecules.

Ionic bonds The attraction between two oppositely charged ions.

Isotopes Sets of atoms that have the same atomic number but different mass numbers.

Ketones Compounds containing a carbonyl group attached between two carbons: $R-\underset{\underset{O}{\|}}{C}-R'$.

Kinetic energy The energy of matter in motion.

Kinetic molecular theory The theory that accounts for the behavior of gases, liquids, and solids by assuming that molecules are in constant, random motion.

LD_{50} The dose of a given substance (in mg of the substance per kg of body weight) that kills 50% of the experimental animals exposed to it.

Law of conservation of mass The total mass of the products produced by a chemical reaction is the same as the total mass of the reactants before the reaction takes place; mass is neither created nor destroyed in a chemical reaction.

Law of constant composition The percent by mass of the elements in a pure compound will always be the same.

Law of multiple proportions If two elements can combine to form more than one compound, the masses of one element which combine with a fixed mass of the other element will always be in that ratio of a small whole number.

Le Chatelier's principle If a system in dynamic equilibrium is subjected to a stress, then the system will change, if possible, to relieve the stress.

Line spectrum The separation of light emitted by gas discharge tubes into a discontinuous, partial rainbow of colors.

Lipids Natural organic materials, such as fats, oils, steroids, and waxes, that dissolve in organic solvents instead of in water.

Liquid The state of matter characterized by fixed volume and arbitrary shape.

Lithosphere All parts of the earth that are neither water nor air: its crust, mantle, and core.

Local anesthesia Complete insensibility to pain in localized areas; one possible effect of a nervous system drug.

London forces The permanent forces of attraction between molecules caused by the cumulative effect of temporary distortions of electron clouds in individual molecules.

Macromolecules Huge molecules containing hundreds or even thousands of atoms.

Main group elements Elements located in the six tallest columns in the periodic table of the elements.

Major energy levels Those allowed regions of space around a nucleus, characterized by a specific amount of energy, where electrons reside.

Mass number The sum of the number of protons and the number of neutrons in the nucleus of an atom.

Mass The amount of matter in a object; its inherent bulk.

Matter Anything that occupies space and has mass.

Measurement Quantitative observation; finding the numerical value of some aspect of matter or energy.

Melting point The temperature at which a solid melts into a liquid; the same temperature as the liquid's freezing point.

Melting The physical change of state from solid to liquid; the opposite of freezing.

Metabolic drugs Drugs that control, supplement, or substitute for various body chemistry processes.

Metabolic poisons Poisons that interfere with normal metabolic processes.

Metabolism A collective term for all of the body processes including digestion, respiration, biosynthesis, and many others.

Metalloids Elements on the borderline (the staircase line on the periodic table of the elements) that have some but not all the properties of a metal.

Metallurgy The series of physical and chemical processes used to obtain a free metal from its ore.

Metals Elements that are shiny, can be pounded into thin sheets, can be drawn into wire, and conduct electricity and heat; elements to the left of and below the staircase line on the periodic table of the elements.

Mineralization The deposit of calcium, phosphate, and hydroxide ions in the form of hydroxyapatite (tooth enamel) on your teeth; the opposite of demineralization or tooth decay.

Minerals Naturally occurring inorganic substances found as various types of rock in the earth's crust; inorganic compounds that are needed in small amounts in the diet.

Moderators Parts of a nuclear fission reactor that slow the neutrons emitted by fission and increase the chances of these neutrons continuing the chain reaction.

Molar mass The mass in grams of one mole of atoms, molecules, or formula unit.

Mole The amount of substance represented by 6.02×10^{23} structural particles of an element or compound.

Molecular formula The representation of the number and kinds of atoms in a molecule.

Monomers Small repeating molecular units that make up a polymer.

Monosaccharides Carbohydrates that contain a single ring; simple sugars.

Mutagens Chemicals that cause mutations.

Mutations Chemical changes in the nitrogen base sequence in a organism's DNA.

Narcotics Addictive drugs that provide relief from intense pain at low doses and produce sleep or stupor at higher doses.

Natural gas Any gas obtained from the ground, but most commonly the gas that contains 50 to 99% methane (CH_4).

Nervous-system drugs Drugs that affect the central nervous system.

Net energy A measure of the true value of any energy source; the total energy of a resource minus the energy needed to find, process, concentrate, and transport it.

Neuron A nerve cell.

Neurotoxins Poisons that disrupt the transmission of nerve impulses.

Neurotransmitters Chemicals released from axons of a nerve cell that trigger a flow of Na^+ and K^+ ions and allow a nerve impulse to travel across a synapse.

Neutral solutions Water solutions in which there are equal amounts of hydroxide (OH^-) ions and hydronium (H_3O^+) ions; neither acidic nor alkaline.

Neutrons Nuclear particles that carry no electrical charge.

Nitro functional group The $-NO_2$ group.

Nitrogen cycle The transfer of nitrogen between living and nonliving things in the environment (Figure 10.4).

Noble gases Elements located in column 18 or VIIIA on the periodic table of the elements.

Nonionic surfactants Surfactants that have no charge on their water-soluble heads.

Nonmetals Elements that lack the properties of metals; elements to the right of and above the staircase line on the periodic table of the elements.

Nonrenewable resource A resource that either is not replaced by natural chemical cycles or is replaced more slowly than it is being depleted.

Nuclear equation The shorthand notation for a nuclear reaction showing the chemical symbol, mass number, and atomic number of all reactants and products.

Nuclear fission A nuclear reaction in which the nuclei of certain isotopes with large mass numbers split apart into two or more isotopes with lower mass numbers.

Nuclear fission reactor A site where fissionable isotopes such as uranium-235 release energy when slow-moving neutrons enter their nuclei and split them into lighter fragments.

Nuclear fuel cycle The transfer of nuclear fuel from mining, to use, to disposal (Figure 11.14).

Nuclear fusion A nuclear reaction at extremely high temperatures in which two or more nuclei with low mass numbers unite or fuse to form a nucleus with a higher mass number.

Nuclear reaction The change of one nucleus to another.

Nucleotides The monomer units of nucleic acids; they consist of a five-carbon sugar, a phosphate unit, and a nitrogen base (Figure 9.18).

Nucleus The tiny, dense core of an atom that contains all of its positive charge and most of its mass.

Octane number A measure of the antiknock properties of a gasoline. The best gasolines have the highest octane numbers.

Octet rule The tendency for atoms to combine such that all the atoms in the compound have eight valence electrons.

Oil shale Underground rock formations that contain a rubbery, solid material consisting mainly of heavy hydrocarbons.

Ores Natural deposits of minerals from which one or more elements may be obtained profitably.

Organic acid A carboxylic acid.

Organic chemistry The study of carbon-containing compounds.

Organic fertilizers Plant nutrients in the form of animal manure, green manure, or compost.

Organic halides Organic compounds containing halogens: fluorine, chlorine, bromine, or iodine.

Organophosphate pesticides Organic chemicals containing phosphorus-oxygen or phosphorus-sulfur functional groups.

Oxidation The loss of one or more electrons or the gain of oxygen in a chemical reaction.

Oxidation-reduction reactions Electron-transfer reactions in which at least one element undergoes oxidation and another undergoes reduction; redox reactions.

Oxidizing agents Compounds that bring about oxidation by acting as electron acceptors and by becoming reduced in the process.

Oxygen cycle The transfer of oxygen between living and nonliving things in the environment (Figure 10.3).

Oxygen debt The shortage of oxygen in muscles that results from bursts of intense work.

Oxygen-demanding wastes Water pollution from organic wastes that use oxygen in their decomposition.

Ozone layer A thin shell of O_3 in the upper atmosphere that absorbs 99% of the sun's harmful ultraviolet radiation.

PANs Peroxyacyl nitrates; components of photochemical smog.

Parkinson's disease A disease characterized by uncontrolled tremors, stiffness, and difficulty in walking; caused by abnormally low concentrations of neurotransmitters.

Parts per million (ppm) A measure of concentration giving the number of dissolved parts in each million parts of the solution; equivalent to the number of milligrams of a substance dissolved in each liter of water.

Passive solar heating Capturing and storing the sun's energy in a specially designed and constructed building.

Penicillin The first antibiotic to be used as a drug; it is produced by the <u>Penicillium</u> mold.

Peptide bond The amide link that connects one amino acid to another in a protein.

Period Each of the seven horizontal rows in the periodic table of the elements.

Periodic table The classification of the elements in rows by increasing atomic number and in columns by similar chemical properties.

Pest Any unwanted organism that directly or indirectly interferes with human activity.

Pesticides Chemicals designed to kill interfering organisms that humans consider undesirable.

Petrochemicals Chemicals made from natural gas, coal, or petroleum.

Petroleum A dark, greenish-brown, foul-smelling liquid containing mostly hydrocarbons; crude oil.

pH scale A measure of the number of hydronium ions in a liter of solution (Figure 6.1).

Pheromones Chemicals that transmit various messages from one organism to another.

Phosphorus cycle The transfer of phosphorus between living and nonliving things in the environment.

Photochemical reactions Any reactions that are activated or catalyzed by light.

Photochemical smog A mixture of primary air pollutants, such as nitrogen oxides and carbon monoxide, that reacts with sunlight to produce secondary air pollutants such as ozone and PANs.

Physical change Any change that does not alter the chemical composition of the substance.

Physical dependence One type of addiction to a chemical; disruption of body functions when the drug is withdrawn.

Plaque Off-white deposits of tooth-decaying bacteria that grow on teeth.

Plasmids Small, freely-floating ringlets of DNA inside most cells that often carry the genetic information for resistance to antibiotics.

Poison Toxic substance; any chemical that causes illness or death in relatively small amounts.

Polar covalent bond A bond with an unequal distribution of charge so that one atom of the bond has a partial negative charge and the other has a partial positive charge.

Polycyclic aromatic compounds Organic compounds with two or more benzene or benzenelike rings fused or joined together.

Polyesters Copolymers formed by the condensation of carboxylic acid and alcohol monomers with repeated ester links.

Polymerization The reaction that links many monomers into a polymer.

Polymers Macromolecules made up of many small repeating molecular units called monomers.

Polysaccharides Carbohydrates that are polymers of the simple sugars or monosaccharides.

Polyunsaturated compounds Compounds with more than one multiple bond.

Positron An electron with a positive charge emitted by a radioactive substance.

Potassium-argon dating A method for finding the age of an artifact by measuring the ratio of potassium-40 to argon-40 it contains.

Potential energy The energy stored in a object because of its position or because of the position of its parts.

Power tower A system of mirrors or lenses to capture the sun's energy; a solar furnace.

Precipitate A solid material formed by a chemical reaction in solution.

Primary air pollutant A harmful chemical that directly enters the air.

Primary sewage treatment The mechanical removal of debris and suspended solids from waste water as a first step in purifying it.

Products The substances that result when reactants undergo a chemical reaction.

Progestogens Female sex hormones that prepare the uterus for pregnancy.

Progression The development stage of cancer caused by joint action of an initiator and a promoter.

Promoter One of the two categories of carcinogens that must act jointly to cause cancer; continued exposure needed after a dose of initiator to develop the cancer.

Proteins Polymers of amino acids.

Protons Nuclear particles that carry the positive charge of an atom.

Psychological dependence One type of addiction to a chemical; habituation; feelings of uneasiness, distress, or anxiety when the drug is withdrawn.

Pure substances Homogeneous matter consisting of the same material throughout; compounds and elements only.

Radioactivity The spontaneous emission of highly energetic rays and particles.

Radiocarbon dating A method for finding the age of an artifact by measuring the amount of carbon-14 it contains.

Radioisotope An isotope of an element whose unstable nuclei are radioactive.

Reactants The starting substances of a chemical reaction that become converted into products.

Reaction rate The quantity of reactants that are converted to products during a specified period of time.

Recombinant DNA DNA from two different sources spliced together to form a new DNA molecule.

Recycling The collection, reprocessing, and refabrication of any product.

Redox reactions Electron-transfer reactions in which at least one element undergoes oxidation and another undergoes reduction; oxidation-reduction reactions.

Reducing agents Compounds that bring about reduction by acting as electron donors and by becoming oxidized in the process.

Reduction The gain of one or more electrons or the loss of oxygen in a chemical reaction.

Rem A measure for the dosage of harmful radiation.

Renewable resource A resource that either is replaced fairly rapidly by natural chemical cycles or comes from an essentially inexhaustible source.

Resource Any form of matter or energy that is obtained from the physical environment to meet human needs.

Restriction enzymes Enzymes that can cut double-stranded DNA at predictable places and leave the DNA with frayed ends containing a segment of just one of the strands.

Reuse Repeated use of a product in its orginal form.

Reverse osmosis A water-purification process in which impure water is forced through membranes whose pores are too small to let the impurities through.

RNA Ribonucleic acid; the single-stranded polymer of nucleotides that participates in protein synthesis in your body.

Rutherford model of the atom The atom is a tiny positively charged nucleus surrounded by empty space through which electrons travel.

Salinity The concentration of dissolved ionic substances in water.

Saponification The reaction between a triglyceride and a strong base to form a soap.

Saturated hydrcarbons Hydrocarbons containing carbon-carbon single bonds only; alkanes.

Schrodinger model of the atom The atom is a complicated entity that can only be described by equations involving higher mathematics; the wave mechanical model.

Scientific data Facts obtained by making observations and measurements in nature.

Scientific hypothesis An educated guess or preliminary idea that explains a scientific law or certain scientific facts.

Scientific law A summary of much scientific data that predicts what will happen in the future under certain conditions.

Scientific method The way scientists gather data and formulate scientific laws, hypotheses, and theories.

Scientific theory A well-tested and widely-accepted scientific hypothesis.

Scrubbers Devices that remove sulfur dioxide from smokestack emissions.

Sebum The oily substance secreted onto your skin and hair.

Second law of energy or thermodynamics Any process increases the entropy of the universe by transforming higher quality energy into lower quality energy.

Secondary air pollutant A harmful chemical that forms in the air because of a chemical reaction between other air components.

Secondary sewage treatment The biological process that uses aerobic bacteria to break down impurities in waste water as a second step in purifying it.

Sedation Calming without producing sleep; one possible effect of a nervous system drug.

Semiconductor A substance, such as a metalloid, with a moderate ability to conduct electricity somewhere between the high conductivity of metals and the low to nonexistent conductivity of nonmetals.

Silicone polymers Polymers made of dihydroxysilicon monomers; siloxanes.

Simple sugars Carbohydrates that contain a single ring; monosaccharides.

Slag A low-melting substance that is produced by the reaction of a flux with an ore.

Soaps The sodium or potassium salts of fatty acids.

Soft water Water that does not contain large concentrations of dissolved mineral ions such as Ca^{2+}, Mg^{2+}, or Fe^{2+}.

Soil erosion The movement of soil, especially topsoil, from one place to another.

Soil sterilant herbicides Chemicals that kill plants by destroying essential microorganisms in the soil.

Soil The thin layer on the earth's crusts consisting of a complex mixture of decaying organic matter, water, air, living organisms, and tiny particles of inorganic minerals and rocks.

Solar cells Photovoltaic cells that convert solar energy directly into electricity.

Solar furnace A system of mirrors or lenses to capture the sun's energy; a power tower.

Solid The state of matter characterized by fixed volume and fixed shape.

Solid waste Any unwanted or discarded material that is not a liquid or a gas.

Solution Two or more pure substances blended together such that all parts of the combination have the same properties and composition; a homogeneous mixture on the molecular level.

Spontaneous process A reaction that proceeds automatically without human help.

Starch A large polysaccharide made of glucose monomers and found in plants such as potatoes, rice, and grains.

Steroids Lipids with a characteristic "chicken wire" carbon frame (Figure 9.10); examples include cholesterol, vitamin D, sex hormones, and cortisone.

Stimulants Drugs that increase the activity of the brain and the central nervous system.

Strong acids Acids that react completely with water to form hydronium (H_3O^+) ions.

Strong bases Bases that release all their hydroxide (OH^-) ions when placed in water.

Structural isomers Two or more compounds that have the same molecular formula but different structures or arrangements of atoms.

Sulfa drugs Any effective synthetic drug that contains an SO_2 group in its structure.

Sunscreens Compounds that absorb ultraviolet light and prevent it from damaging the skin.

Surface tension The force that causes the surface of a liquid to take the smallest possible area.

Surface water That liquid water on earth that exists above ground in rivers, oceans, wetlands, and other places.

Surfactants Surface-active agents; molecules with a water-soluble, oil-repellent head and an oil-soluble, water-repellent tail.

Surroundings All the other matter in the universe besides the system or matter under study.

Synapse The small gap between the axon of one nerve cell and a dendrite of another.

Synfuels Synthetic natural gas or oil obtained from coal.

Synthetic laundry detergents Mixtures of surfactants, builders, and other laundering aids.

System The collection of matter under study, as opposed to the surroundings.

Systemic herbicides Chemicals that kill plants by causing excess production of growth hormones.

Tar sands Deposits that contain a black, high-sulfur, tarlike oil known as bitumen; oil sands.

Technology The means used for making of new products and processes to improve our survival, comfort, and quality of life; applied science.

Teratogens Chemicals that cause birth defects.

Tertiary sewage treatment A series of specialized chemical and physical processes that remove pollutants left by the primary and secondary treatment.

Tetraethyl lead A fuel additive used to increase the octane rating of lower quality gasoline and a major source of lead pollution in the atmosphere; $(CH_3CH_2)_4Pb$.

Thermal enrichment An increase in water temperature that proves to be beneficial to the environment of interest.

Thermal inversion A recurring weather condition characterized by a layer of dense, cool air trapped under a layer of light, warm air in an urban area or valley; temperature inversion.

Thermal pollution An increase in water temperature that proves to be harmful to the environment of interest.

Thermoplastic polymers Polymers that can be softened and molded into almost any shape.

Thermosetting polymers Polymers that become permanently hard and rigid once they have been melted.

Thomson model of the atom The atom is a positively charged sphere with electrons imbedded in it like seeds in a watermelon.

Tranquilizers Drugs that depress the activity of the brain and the central nervous system.

Triglycerides Compounds made when three fatty acids form an ester with glycerol.

Troposphere That part of the atmosphere, from 0 to 12 km above the earth's surface, that contains most air pollution.

Unsaturated hydrocarbons Hydrocarbons containing double or triple bonds between any two carbon atoms.

Urban heat island A dome of heat over a city created by the slow release of energy from concrete and brick.

Vaccines Injections of weakened forms of harmful bacteria that help the immune system build antibodies against them.

Valence electrons Electrons in the outermost occupied main energy level of an atom; the electrons that participate in chemical bonds.

Vapor pressure The pressure of the gas or vapor at equilibrium over a liquid in a closed container.

Vaporization The evaporation of a liquid below its boiling point; the conversion from the liquid state to the gaseous.

Vinyl polymers Polymers made from the vinyl monomer, $CH_2=CHR$.

Vitamins A group of some 21 organic compounds that are needed in small amounts in the diet for good health.

Volume The amount of space an object occupies.

Water pollution The degradation of water by a substance or a condition (such as heat) to the extent that it cannot meet water quality standards or cannot be used for a specific purpose.

Wave mechanical model of the atom The atom is a complicated entity that can only be described by equations involving higher mathematics; the Schrodinger model.

Weak acids Acids that react only slightly with water to form hydronium (H_3O^+) ions.

Weak bases Bases that are either react only slightly with water to produce hydroxide (OH^-) ions or are not soluble enough in water to produce many.

Weight The gravitational pull on an object.

Wetting The coating of a surface with a layer of liquid.

Wind energy The kinetic energy of the moving atmosphere.

Urban heat island: [illegible] over [illegible] created by [illegible]
[illegible]

Vacuum [illegible] [illegible]
[illegible] system [illegible]

Valence electron: [illegible] electrons [illegible]
[illegible]
[illegible] bonds.

Vapor pressure: The pressure [illegible]
[illegible]

Vaporization: [illegible]
[illegible]

[illegible]

[illegible]
[illegible]

Volume: the amount of [illegible]

[illegible]
[illegible]
[illegible]

[illegible]
[illegible]
[illegible]

[illegible]
[illegible]

[illegible]
[illegible]
[illegible]

[illegible]

[illegible]

[illegible]